BIOSTATISTICS
FOR POPULATION
HEALTH

A PRIMER

LISA M. SULLIVAN, PhD
Associate Dean for Education, Professor of Biostatistics,
Boston University School of Public Health

JONES & BARTLETT
LEARNING

World Headquarters
Jones & Bartlett Learning
5 Wall Street
Burlington, MA 01803
978-443-5000
info@jblearning.com
www.jblearning.com

Jones & Bartlett Learning books and products are available through most bookstores and online booksellers. To contact Jones & Bartlett Learning directly, call 800-832-0034, fax 978-443-8000, or visit our website, www.jblearning.com.

34820-0

Production Credits
VP, Product Management: Amanda Martin
Director of Product Management: Laura Pagluica
Product Manager: Sophie Fleck Teague
Product Specialist: Sara Bempkins
Project Manager: Jessica deMartin
Project Specialist: David Wile
Digital Project Specialist: Rachel DiMaggio
Senior Marketing Manager: Susanne Walker
Manufacturing and Inventory Control Supervisor:
 Therese Connell
Composition: codeMantra U.S. LLC
Project Management: codeMantra U.S. LLC
Cover Design: Kristin E. Parker
Text Design: Kristin E. Parker
Senior Media Development Editor: Troy Liston
Rights & Media Specialist: Maria Leon Maimone
Cover Image (Title Page, Chapter Opener):
 © MicroStockHub/iStock/Getty Images Plus/Getty
 Images; © Zenza Flarini/Shutterstock
Printing and Binding: McNaughton & Gunn

Library of Congress Cataloging-in-Publication Data
Names: Sullivan, Lisa M. (Lisa Marie), 1961- author.
Title: Biostatistics for population health: a primer / Lisa Sullivan.
Description: Burlington, MA: Jones & Bartlett Learning, [2021] | Includes bibliographical references and index. |
 Summary: "This short format primer will provide foundational coverage of biostatistical concepts and applications for health professionals. It will use examples that are relevant for clinical and health professionals specifically"—Provided by publisher.
Identifiers: LCCN 2019051160 | ISBN 9781284194265 (paperback)
Subjects: MESH: Biostatistics—methods | Population Health—statistics & numerical data | Statistics as Topic
Classification: LCC QH323.5 | NLM WA 950 | DDC 570.1/5195—dc23
LC record available at https://lccn.loc.gov/2019051160

6048

Printed in the United States of America
24 23 22 21 20   10 9 8 7 6 5 4 3 2 1

Contents

Acknowledgment

I am grateful for the support of so many people. First, many thanks to Dr. Richard Riegelman for his unending enthusiasm, vision and valuable feedback. Thank you to Kevin Green, Nina Huttemann, Ryan Sullivan and Kim Dukes for their careful review and suggestions for improvement. I am forever thankful to so many extraordinary students and colleagues with whom I have the privilege to work at the Boston University School of Public Health. And finally, thank you to my family and friends – who support and inspire me every day.

Introduction

Improving the health of populations takes coordinated efforts that address individual, social, environmental, economic, and political factors. Data are key in improving population health. We need data to describe the burden of disease, quantify associations between risk factors and outcomes, shine a light on inequities in health outcomes, and evaluate whether treatments and interventions are effective in reducing disease and promoting health. But to be useful, data must be relevant, carefully collected, managed, analyzed, and interpreted.

In this primer, we review approaches and techniques in applied biostatistics that are regularly used by health professionals to turn data into knowledge for action. We illustrate computations using simple, but realistic, examples and paying special attention to appropriate interpretation of biostatistical results.

The primer is organized into three units:

1 Summarizing data for decision making
2 Associations between two variables
3 Multivariable analysis

In Unit 1, we discuss techniques for summarizing data for decision making. We begin by defining different types of variables (also called data elements) and how they are measured. We then discuss popular methods for summarizing different types of variables. We define and contrast rates, proportions, and ratios and how they are used in population health studies. We conclude with a brief discussion of graphical displays of data, which are particularly powerful in translating biostatistical results to both scientific and lay audiences.

In Unit 2, we discuss associations between two variables. In many biostatistical applications, we are interested in the association between a risk factor and an outcome. Is a particular behavior associated with a better health outcome? Does a particular policy adversely affect health outcomes? Are individuals with a particular genetic makeup more susceptible to disease? Is a new treatment safe and effective?

We begin by reviewing key concepts in probability as these underpin the techniques we use to quantify associations. We then discuss screening and diagnostic tests and, more specifically, how to assess the performance of screening tests for improving health. We then move on to statistical inference, where we apply biostatistical techniques to make inferences about populations based on the analysis of samples. We focus on appropriate interpretation of statistical results, recognizing uncertainty and avoiding overstating the results, which often occur in the translation of statistical findings. We discuss confidence interval estimates and procedures for hypothesis testing in detail.

In Unit 3, we discuss multivariable analysis, moving beyond the association between two variables to take into account relationships among other factors that are inevitably at play. We discuss techniques to handle confounding (i.e., a distortion [exaggeration or masking] of an association between a predictor and an outcome due to another variable) using popular statistical applications such as multiple linear regression analysis and multiple logistic regression analysis. We also briefly discuss survival analysis techniques, again focusing intentionally on appropriate interpretation of results.

Each unit is organized in the same way and begins with a current population health issue. Framing questions are articulated that exemplify questions of importance and impact that come up in practice and where data and statistical analysis can be used to motivate action. Biostatistical techniques are then described and illustrated. Practical examples are interspersed to engage health professionals in building skills, intuition, and confidence. Each unit ends with a brief summary of key points covered.

UNIT 1

Summarizing Data for Decision Making

D ata are important for decision making in population health. Data can be extremely powerful and convincing when used appropriately. Misunderstanding and misapplications, however, can lead to incorrect conclusions and recommendations. To be most useful, data must be relevant and taken in context, and that context must be well understood to use data effectively. In this unit, we review different types of data, distinct data collection strategies, and measurement issues that can arise. We then review popular methods to summarize or describe data. We discuss popular measures to compare groups, focusing on risks, rates, and ratios. And, finally, we discuss graphical displays of data as powerful tools for action. With each topic, we highlight issues of context that need careful consideration.

A Population Health Issue—The Opioid Crisis in the United States

As of March 2018, the National Institutes of Health reports that more than 115 people in the United States die every day of opioid overdose, more than 20% of people who are prescribed opioids for chronic pain misuse them, about 5% of people who misuse opioids transition to heroin, and approximately 10% of people who are prescribed opioids develop an opioid use disorder.[1]

In this unit, we discuss biostatistical techniques that will allow us to answer questions such as the following:

- What are the characteristics of adolescents who misuse opioids?
- Do overdose deaths vary across states in the United States?
- What statistics and data visualizations might be effective to inform adolescents about the risks of opioid misuse?

Before we delve into the specifics of summarizing data, let us first discuss the power of statistics.

Populations and Samples

We are often interested in making generalizations about specific attributes, behaviors, and outcomes in populations. For example, what percentage of patients undergoing total knee replacement surgery develop an infection in the month following surgery? Is one glass of red wine per day healthy or harmful? Are there safe and effective treatments for chronic pain? Are women who smoke during pregnancy at higher risk for premature delivery?

Let us consider the last question: Are women who smoke in pregnancy at higher risk for premature delivery? To answer this question definitively requires an analysis of all pregnancies by assessing smoking status and outcomes of each pregnancy. Even if we restricted our attention to a reasonably small geographic area at a particular time, this would be a practically impossible task. In fact, it is not necessary to study all pregnancies to precisely understand the link between smoking in pregnancy and premature delivery. This is where statistics, and specifically the power of statistics, comes in.

A collection of all individuals with a particular condition or who possess a specific attribute make up a population of interest. In most situations, the population is very large, making it prohibitive, financially and practically, to involve every member of that population in an analysis. The population size is usually referenced by N (e.g., $N = 100,000$ or $N = 1,000,000$). In practice, we take a sample of individuals from the population, where the sample size is denoted by n (e.g., $n = 125$, $n = 75$). The sample is a subset, smaller in size (i.e., fewer individuals), and ideally representative[a] of the population (there are specific techniques for selecting samples from populations—please see Lohr (2010) for details).[2] In practice, we analyze samples. If the sample is representative of the population, then what we observe in the sample should generalize to the population. This is the power of statistics. We make inferences about a large population based on a study of only a fraction of the membership of that population (i.e., our analysis sample). However, it is a big leap from a sample to a population and there are many issues that need careful consideration.

First and foremost, the sample we analyze on any issue is merely one sample of many that might have been possible. This is critical to bear in mind, as a different sample might produce different results. When we make generalizations or inferences about a population based on what we observe in one sample, we build in an estimate of sampling variability. While we can never be certain about a population based on one sample, we can often generate estimates about populations that are sufficiently precise for decision making. We will discuss this in more detail in Unit 2.

---

a By representative, we mean that the participants in the sample "look like" members of the population in terms of important demographic, social, and clinical characteristics or attributes.

We now return to context, and how data must be considered in context to be useful. **EXAMPLE 1·1** frames some of the issues we must consider in statistical analysis.

EXAMPLE 1·1

Suppose we wish to compare two hospitals in terms of the likelihood that their patients develop infection following total knee replacement surgery. We focus our analysis on infections that are detected in the month following surgery, and we consider surgeries performed between January 1, 2018, and December 31, 2018. Suppose that 10 patients who had surgery in hospital A developed infection compared to 5 patients who had surgery in hospital B. Twice as many patients in hospital A compared to hospital B developed infection in 2018. Is it fair to conclude that patients undergoing surgery in hospital A are at twice the risk for infection following total knee replacement surgery?

Let us dig a bit deeper. Suppose that hospital A performed far more total knee replacement surgeries than did hospital B in 2018. Is the volume of surgeries important to consider? Suppose that hospital A performed 1,100 total knee replacement surgeries in 2018 compared to 245 total knee replacement surgeries performed in hospital B. The proportion of patients in hospital A who developed infection in 2018 was 10/1,100 = 0.0091, or less than 1%. The proportion of patients in hospital B who developed infection in 2018 was 5/245 = 0.0204, about 2%, which is twice the proportion in hospital A. The counts (10 vs. 5) are important but insufficient to make a comparison between hospitals A and B in terms of infection following total knee replacement surgery. Studies have also shown that patients with diabetes are at higher risk of infection following total knee replacement surgery.[3] What if hospital A had more patients with diabetes? Is that important to consider? How might that affect our comparison?

▶ 1.1 Data, Measurement, and Variables

To truly understand a particular attribute, behavior, or outcome, we must be able to measure it. There are multiple methods to measure data that are relevant in clinical and public health practice. These include self-administered or interviewer-administered surveys to collect data from participants or patients, the use of medical devices that take measurements directly, and the extraction of data from electronic health records that contain extensive clinical data. The appropriate tools and techniques to gather data depend on what is being measured. Above all, we aim to measure accurate or valid (i.e., correct) data.

In most physical examinations, vital signs and other important information are gathered. Vital signs are measured according to specific protocols so that comparisons can be made over time in the same patient and among patients. The units of measurement are always important to note when measuring and summarizing data. For example, pulse rate is measured by counting the number of times the heart beats per minute—the "pulses"—in the radial, brachial, or carotid arteries over a specified period. Body temperature is measured in degrees (Fahrenheit or Celsius) and can be captured using a glass or digital thermometer, and blood pressure is measured in millimeters of mercury (mmHg) using a sphygmomanometer. It is always important to convey units of measurement when reporting data and statistical results.

We use the term "variable" to reflect any phenomenon, characteristic, or attribute of interest. A variable might reflect a particular characteristic (e.g., patient age, race/ethnicity, family history), behavior (e.g., smoking, physical activity), laboratory measurement (e.g., total serum cholesterol), condition (e.g., pregnancy), or diagnosis (e.g., diabetes). We must also classify variables as risk factors or predictors compared to outcomes or responses. Regardless of the role a variable might play in an analysis (risk factor or outcome), the first step in any analysis is to classify key variables by type. Most analyses involve variables that are one of the following four types: dichotomous, categorical, ordinal, or continuous. It is very important to correctly classify variables, as different statistical techniques are used for the different variable types. Definitions of each variable type along with a few examples clarify this point:

Dichotomous Variables

Also called binary variables, dichotomous variables have two possible response options (e.g., yes/no). Examples of dichotomous variables include presence of infection, diagnosis of Type I diabetes, biological sex at birth, and currently taking antihypertensive medications.

Categorical Variables

Also called nominal variables, categorical variables have two or more unordered response options (note that dichotomous variables can also be called categorical variables with two response options, e.g., yes/no). Examples of categorical variables include race/ethnicity, eye color, and cancer type (e.g., breast cancer, colorectal cancer, gastric cancer, melanoma).

Ordinal Variables

Ordinal variables have two or more ordered response options (note that dichotomous variables can also be called ordinal variables with two ordered response options, e.g., low/high). Examples of ordinal variables include cancer stage (I, II, III, IV), blood pressure category (optimal, normal, prehypertensive, hypertensive), and self-reported health status (excellent, very good, good, fair, poor).

Continuous Variables

Sometimes called measurement or quantitative variables, continuous variables take on any value within a realistic or reasonable range. Examples of continuous variables include age in years, weight in pounds, height in inches.

In the next section, we describe and illustrate techniques to summarize each variable type.

▶ 1.2 Descriptive Statistics

Descriptive statistics include any summary measures computed on a sample. Here we discuss a number of popular descriptive statistics, organized by variable type. In most analyses, investigators have many variables that they are analyzing simultaneously. Summary tables or reports include various descriptive statistics that are tailored to specific variable types.

In every analysis, we report the sample size, or number of individuals in that sample. The sample size is denoted by n (e.g., $n = 50$, $n = 125$, $n = 10,453$). Generally speaking, larger samples produce more precise statistical results. However, in some studies it is impossible to sample large numbers as we may be investigating a rare disease or a risk factor or outcome that occurs infrequently. It is also true that there is a point at which more individuals are not needed to increase precision. Careful consideration must always be given to determine the sample size needed to ensure that the estimates from the sample are useful for decision making.

EXAMPLE 1-2 is an application that involves variables of differing types that will be summarized using the descriptive statistics we present here.

EXAMPLE 1-2

Suppose we want to understand the reasons why adults between the ages of 18 and 64 use an emergency department (ED). A survey is designed to capture the primary reason for an ED visit, such as lack of access to a primary care medical practitioner, the primary care medical practitioner not being available, the seriousness of the problem, and so on. The survey is given to each adult visiting the ED while in the waiting room. The survey also captures demographic and medical history data. Suppose that investigators have the time and resources to collect data on 150 patients (i.e., $n = 150$). Once all the data are collected, investigators plan to summarize the patients in their sample using a table similar to **TABLE 1-1**. The primary focus of the survey is to understand the reasons why adults use EDs, but it is important to understand the sample to interpret these reasons in context.

(continues)

EXAMPLE 1-2 *(continued)*

TABLE 1-1 Table to Summarize Demographic Characteristics of $n = 150$ Adults Using an Emergency Department

Demographic Characteristic	Descriptive Statistics
Age (years)	
Male sex	
Race/Ethnicity	
Black	
Hispanic	
White	
Other	
Educational Attainment	
Less than high school graduate	
High school graduate	
Some college or college graduate	
Some graduate or graduate degree	
Currently have medical insurance coverage	

Table 1-1 contains a mix of variable types, and all are important to convey the demographic composition of the sample. Specifically, age is continuous, male sex is an indicator of biological sex at birth and is dichotomous, race/ethnicity is categorical, educational attainment is ordinal, and current medical insurance coverage is dichotomous. The appropriate descriptive statistics to report in the right column of Table 1-1 depend on the variable type.

Next we outline descriptive statistics for each variable type, illustrate the computations with small samples (so as to not get too bogged down in computations), and then return to our example to complete Table 1-1 for our sample of $n = 150$ adults (see Table 1-4).

Descriptive Statistics for Dichotomous Variables

Dichotomous variables have two response options. The most appropriate descriptive statistics for dichotomous variables are the numbers and percentages of respondents in each response category, also called the frequencies and relative frequencies, respectively. With dichotomous variables it is only necessary to report the number and percentage in one group as the remainders are assumed to fall into the other group, assuming there are no missing data or individuals in the sample who did not respond or provide data.

EXAMPLE 1-3 illustrates the approach.

EXAMPLE 1-3

Suppose we sample $n = 10$ patients seeking primary care at a particular ambulatory care center and ask each whether they have taken prescription pain pills during the past 6 months. Each patient responds either "yes" or "no." For analytic purposes, dichotomous variables are often coded 0 for no and 1 for yes. The sample data are below, coded $0 = $ no and $1 = $ yes, for each of the $n = 10$ participants in the sample.

$$0 \quad 1 \quad 0 \quad 0 \quad 1 \quad 1 \quad 0 \quad 0 \quad 0 \quad 0$$

An appropriate summary would be that three (30%) of $n = 10$ participants report taking prescription pain pills during the past 6 months.

Descriptive Statistics for Categorical and Ordinal Variables

Categorical and ordinal variables have two or more unordered and ordered response options, respectively. Again, the most appropriate descriptive statistics are the numbers (frequencies) and percentages (relative frequencies) of respondents in each response category. With more than two response options it is important to show the number and percentage in each group.

Continuing with the small sample of $n = 10$ patients in Example 1-3, suppose we ask each to report their perceived mental health status on the following scale: very good, good, fair, poor. Note that this is an example of an ordinal variable and the sample data below are for each of the $n = 10$ participants in the sample:

| very good | very good | poor | poor | fair |
| good | good | fair | good | very good |

An appropriate summary of this data is shown in **TABLE 1-2**.

TABLE 1-2 Self-Reported Mental Health Status in Sample of $n = 10$ Patients Seeking Primary Care

Self-Reported Mental Health	n (%)
Very good	3 (30)
Good	3 (30)
Fair	2 (20)
Poor	2 (20)

Descriptive Statistics for Continuous or Measurement Variables

Continuous variables take on any value within a realistic or reasonable range. There are several popular descriptive statistics for continuous variables, and very broadly, they describe central tendency and variability. There are several different measures of central tendency and of variability—the best summary measures to report in a specific application depend on whether the variable is subject to outliers (extremes) or not.

Central tendency is often described by the sample mean, $\bar{X}$. The sample mean is computed by adding up all values and dividing the total by the sample size (i.e., $\bar{X} = \dfrac{\Sigma x}{n}$, where x denotes the variable of interest).

Consider again our small sample of $n = 10$ participants in Example 1-3 and suppose we measure weight in pounds, a continuous variable, and observe the following:

150	125	175	190	210	225	170	140	190	200

The sample mean is computed by adding up the weights ($\Sigma x = 1775$) and dividing the total by the sample size: $\bar{X} = \dfrac{\Sigma x}{n} = \dfrac{1775}{10} = 177.5$ pounds. Thus, the mean or typical weight in the sample is 177.5 pounds.

Another measure of central tendency is the median. The median is the middle value. The median separates the highest 50% of the measurements from the lowest 50% and is used to summarize central tendency when a variable is subject to outliers. In our small sample of $n = 10$ participants, the median is computed by first ordering the measurements (weights) from lowest to highest (or highest to lowest).

125	140	150	170	175	190	190	200	210	225

Because there is an even number of measurements ($n = 10$), the median is the mean of the two middle weights $(175 + 190)/2 = 182.5$ pounds.

125	140	150	170	175		190	190	200	210	225

(50% of weights below median) ↑ (50% of weights above median)

median = 182.5 pounds

In this small sample there are no obvious outliers. When the mean and median are close in value (e.g., the mean weight is 177.5 pounds and the median weight is 182.5 pounds), we can assume that there are no extremes and can report the mean. By way of example, triglyceride levels in blood, often measured as part of a lipid panel, might have a mean of 150 mg/dL in a sample of adults who are free of cardiovascular disease (CVD) and might be higher in adults with heart disease. Triglycerides can vary widely. Some people have high triglyceride levels (in the range of 200–499 mg/dL), while others may have very high triglyceride levels (exceeding 500 mg/dL). If most participants in a sample have triglycerides below 200 mg/dL and one or two have values exceeding 500 mg/dL, the mean will be inflated toward the extreme (or outlying) values and the median is not affected by these extremes. In the presence of outliers, the median is a better representation of central tendency.

Variability is crudely measured by the absolute range of a continuous variable. The absolute range is the difference between the minimum and maximum values observed. Sometimes people report the range as this difference or report the range as between the minimum and maximum value. In our small sample of $n = 10$ weights, the minimum recorded weight is 125 pounds and the maximum is 225 pounds. The range can be reported as either 100 pounds or between 125 and 225 pounds.

A more popular measure of variability is the sample standard deviation, s. The standard deviation is a measure of how far each observed value is from the sample mean. Larger standard deviations indicate more variability among measurements in the sample; smaller standard deviations indicate more clustering of measurements around the sample mean. The sample standard deviation is computed by taking the difference between each measurement and the sample mean, and then summarizing the differences. The standard deviation is defined as $s = \sqrt{\dfrac{\Sigma(x - \bar{X})^2}{(n-1)}}$. In our sample of $n = 10$ weights, we first compute the difference between each weight and the sample mean weight $\left(\bar{X} = 177.5 \text{ pounds}\right)$. We then square these differences (so as to capture the magnitude of the differences without the negative and positive differences cancelling each other out), add up the differences, divide by $(n - 1)$, and take the square root.[b] The steps are illustrated with our small sample.

TABLE 1-3 displays the computations for the sample standard deviation, s, in weights.

---

b When computing the sample standard deviation, s, we divide the sum of the squared differences by $(n - 1)$ instead of n. The reason for this is if we try to estimate the true unknown population variance based on the sample variance and compute the sample variance by dividing by n, we underestimate the true population variance. However, if we divide by $(n - 1)$, we get what is called an unbiased estimate of the true population variance.

TABLE 1-3 Calculating the Sample Standard Deviation, s

Observed Weights, (x)	Differences from Mean Weight, $(x - \bar{X})$	Differences from Mean Weight Squared, $(x - \bar{X})^2$
150	−27.5	756.25
125	−52.5	2,756.25
175	−2.5	6.25
190	12.5	156.25
210	32.5	1,056.25
325	47.5	2,256.25
170	−7.5	56.25
140	−37.5	1,406.25
190	12.5	156.25
200	12.5	156.25
		$\Sigma(x - \bar{X})^2 = 8{,}762.5$

The sample standard deviation is $s = \sqrt{\dfrac{8{,}762.5}{9}} = \sqrt{973.61} = 31.2$. Variability in weights are summarized as follows. In this sample, individuals' weights are approximately 31.2 pounds from the mean weight of 177.5 pounds.

Another often-reported statistic is the standard error. The standard deviation, s, quantifies variability among observations within a sample. The standard error quantifies variability in a summary statistic (e.g., the sample mean) among samples. The standard error of the sample mean is defined as $s/\sqrt{n}$.[c] In practice, we have one sample and we compute summary statistics (e.g., n, $\bar{X}$, s) on that sample. But that sample was one of many possible samples, and if we had selected different participants into our sample, we might have observed a different sample mean (or any other summary statistic).

---

c Note that the standard error, $s/\sqrt{n}$, is inversely related to the sample size, n. With larger samples we have less variability in estimates of the sample mean (i.e., less sampling variability).

The standard error is a measure of sampling variability. It gives a sense of how variable estimates might be from different samples. Quantifying precision (or lack thereof) in statistical estimates is a critical aspect of statistical analysis. We discuss this in detail in Unit 2.

When a measurement is not subject to extreme or outlying values, the sample mean and sample standard deviation are appropriate measures of central tendency and variability, respectively. When a measurement is subject to extremes (e.g., triglycerides), the median is a more appropriate measure of central tendency. The interquartile range (IQR) is the most appropriate measure of variability when a measure is subject to extremes or outliers. The IQR is the difference between the third quartile (defined as the value that separates the top 25% of the measurements from the rest) and the first quartile (defined as the value that separates the bottom 25% of the measurements from the rest). Similar to the range (which can be reported as the difference between the maximum and minimum values or as from the minimum to the maximum value), the IQR can be reported either as the difference between the first and the third quartile or as from the first to the third quartile.

Returning to our small sample of $n = 10$ participants, the median weight is computed by first ordering the observations from smallest to largest and then computing the mean of the two middle weights, $(175 + 190)/2 = 182.5$ pounds. Again, the median separates the top 50% of weights from the bottom 50% of weights. The first quartile is the weight that separates the bottom 25% of the weights from the rest. We can think of the first quartile as the middle of the lower 50% of the weights, in this case 150 pounds. The third quartile is the weight that separates the top 25% of the weights from the rest, or the middle of the upper 50% of the weights, in this case 200 pounds.

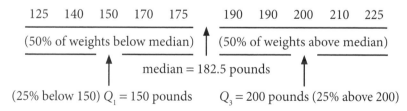

Our small sample of weights is best summarized with the mean and standard deviation—$\bar{X} = 177.5$ pounds and $s = 31.2$ pounds—because there are no outliers. If the sample had outliers or extremes (e.g., if the maximum weight was 425 pounds instead of 225 pounds), then the median and IQR would be more appropriate: median = 182.5 pounds, IQR = 50 pounds (or IQR from 150 to 200 pounds).

Returning to our more realistic sample of $n = 150$ adults using the ED, we would very likely compute descriptive or summary statistics using a statistical computing package rather than manually. The data are analyzed using a statistical computing package and summarized in **TABLE 1-4**.

TABLE 1-4 Demographic Characteristics of $n = 150$ Adults Using an Emergency Department

Demographic Characteristics	$\bar{X}(s)$ or n (%)
Age (years)	42.4 (6.2)
Male sex	87 (58)
Race/Ethnicity	
Black	42 (28)
Hispanic	38 (25)
White	54 (36)
Other	16 (11)
Educational Attainment	
Less than high school graduate	24 (16)
High school graduate	65 (43)
Some college or college graduate	33 (22)
Some graduate or graduate degree	28 (19)
Currently have medical insurance coverage	22 (14.7)

We might describe the sample in a report or a paper as 58% male with a mean age of 42.4 years and a standard deviation of 6.2 years. The patients using the ED are racially diverse, with slightly more white (36%) compared to black (28%) or Hispanic patients (25%) using the ED. Nearly 20% of patients using the ED have some form of graduate education, another 22% have some college education, and 16% never graduated from high school. Only 15% currently have medical insurance. Most research papers and study reports include a table like Table 1-4 to provide the reader with a sense of the participants in the sample. It is extremely important to understand the participants in any sample because when we generalize results from a sample to a larger population, we can only make inferences about the population from which the sample was drawn.

▶ 1.3 Risks, Rates, and Ratios and Their Use in Population Health

In the previous section, we discussed different variable types and how best to summarize each type. Dichotomous variables might seem the simplest to measure and summarize, and in some ways this is true, but there are nuances that need consideration in the summary and interpretation of dichotomous variables. Many applications in population health involve dichotomous risk factors (e.g., a person might, or might not, be exposed to a particular occupational or environmental hazard, engaged, or not, in a particular unhealthy behavior, or following a specific treatment protocol or not) and dichotomous outcomes (e.g., a patient might, or might not, suffer an injury, develop disease, or die within a particular period). In the previous section, we focused on summarizing dichotomous variables with counts (frequencies) and proportions or percentages (relative frequencies). Counts of individuals exposed to a certain risk factor, who engage in a particular behavior, or who develop an outcome of interest are very important. However, counts alone tell us little about population risk for certain exposures and outcomes, or the rate and over what time period disease develops. Here we define and illustrate the use of risks, rates, and ratios in population health applications.

In the previous section, we computed the proportion of males in our study sample of $n = 150$ by $87/150 = 0.58$, or 58%. Proportions are computed by taking the ratio of a part to a whole, the numerator is a subset of the denominator, and proportions are always between 0 and 1. Dichotomous variables are summarized by counts and proportions, and proportions can be converted into percentages by multiplying by 100, which are often easier to interpret, particularly by lay audiences.

Proportions and percentages can be used to describe things such as the proportion of patients who fail to show up for scheduled clinic visits, the percentage of pediatric cancer patients with leukemia, the proportion of Hispanic children who meet the criteria for obesity, the percentage of first graders in the United States with learning disabilities, and so on. Proportions and percentages are widely used in population health, and it is vital to understand the difference, percentage = proportion × 100. There are a number of important issues that need careful consideration so that they are computed and interpreted appropriately. **EXAMPLE 1-4** illustrates the approach.

EXAMPLE 1-4

Suppose we want to summarize the burden of CVD in women over the age of 50 living in Boston in 2018. Note how specifically we define our question. Highly specific questions are best from an analytic standpoint. Thus, we should strive to be explicit and specific in each question we pose. Based on 2017 census data, there were approximately 100,000 women over the age of 50 living in Boston.[4] It would be practically impossible to survey each woman to answer the question of interest. Suppose instead that we have the resources to sample $n = 500$ women. Because we need to assess whether each woman had been diagnosed with CVD, we focus on women who receive medical care at one of the four large medical centers in Boston (recognizing that we may miss cases of CVD among women not receiving medical care). For each woman selected, we review her electronic medical record to assess whether she had been diagnosed with CVD as of December 31, 2018. Among the

(continues)

EXAMPLE 1-4   (*continued*)

$n = 500$ women sampled, we found that 72 of the women had diagnoses of CVD in their electronic medical records. Thus, we report that 72 (14.4%) of $n = 500$ women over the age of 50 living in Boston in 2018 have documented CVD. This percentage is an estimate of the prevalence of CVD. Prevalence refers to the number of existing cases of a disease as of a particular time (in this case, CVD in women over 50 years of age living in Boston in 2018).

The estimate of prevalence is important, but it does not address how long the women have had disease or the rate at which women might develop CVD. The latter is called incidence of disease and reflects new as opposed to existing (prevalent) cases. Prevalence and incidence are both important measures, and we must understand what each conveys and how each of these measures might be used for decision making. Prevalence can be used to describe the burden of an exposure, risk factor, or outcome in a population. Incidence can be used to describe the likelihood, chance, or risk that a patient will develop disease over a specified time period. We will discuss incidence, also called risk, in detail.

Risks

In order to estimate incidence (also called the incidence proportion, or risk), we sample participants who are free of the outcome of interest. We then specify a time frame over which we follow each individual for the development of that outcome. Returning to Example 1-4, suppose we sample a different group of $n = 500$ women from our four medical centers who were over the age of 50 and free of CVD as of January 1, 2018. We review each woman's electronic medical record for a diagnosis of CVD over the next 12 months and record 18 new cases of CVD during 2018. Our estimate of the incidence of CVD is $18/500 = 0.036$ per year, or 36 new cases per 1,000 women per year. When reporting incidence, we often convert the proportion (computed by dividing the number of new cases by the total number of persons at risk) into an integer by multiplying by 100, 1,000, or whatever multiplier it takes to produce a more interpretable result.

Incidence is often reported as annual incidence, and our previous example was set up to measure new cases over a 1-year follow-up period, translating readily to an estimate of annual incidence. Sometimes we follow participants in studies for development of disease over longer periods of time. This is particularly true for diseases with long latency. For example, suppose that in the previous study we had recruited $n = 500$ women into the study who were over the age of 50 and free of CVD as of January 1, 2014, and followed them for 5 years for development of CVD. Suppose that we use the same protocol and again review each woman's electronic medical record for a diagnosis of CVD over the 5 years following enrollment and record 81 new cases of CVD from January 1, 2014 through December 31, 2018. The estimate of incidence of CVD is 81/500 = 0.162 over 5 years, or 16.2 new cases per 100 women (or 162 new cases per 1,000 women) over 5 years. This translates to an annual incidence of 3.2 new cases per 100 women per year. (Note that the estimate of annual incidence based on the 5-year study is not identical to that based on the 1-year study due to sampling variability, which we will discuss in more detail in Unit 2.)

Both prevalence and incidence proportion (risk) are computed as proportions. Often prevalence is reported as a percentage; for example, as of December 31, 2018, 14.4% of women 50 years of age and older living in Boston have CVD. The incidence proportion, or risk, is also computed as a proportion and often reported as incidence per year, or annual incidence. Based on Example 1-4, just over 3 women per 100 in this age group are expected to develop CVD annually. Another interpretation of this is just over 3% of women over age 50 living in Boston are at risk to develop CVD per year.

Rates

In some studies, we aim to estimate incidence or risk of developing certain outcomes that have long latency and we monitor participants over long periods of time. Longer follow-up periods are often necessary to observe a sufficient number of cases to ensure precision in statistical estimates. However, there are challenges in studies with a longer follow-up time. For example, we might lose people over time (e.g., they move away from the area, drop out of the study for one reason or another, become disinterested), or they might die due to a cause unrelated to the outcome of interest. In such situations it is not possible to compute the incidence proportion directly because we do not know the outcome status for each participant at risk (i.e., we do not know whether they develop, or do not develop, the outcome of interest over the follow-up period). For participants who do develop the outcome of interest, we know the time at which they develop the outcome. For those who do not develop the outcome of interest, we know only the last time when they were outcome-free. Thus, we cannot compute the incidence proportion because we do not have complete information on every participant. Instead, we use all observed data (never discarding data) to compute an incidence rate. An incidence rate is computed by dividing the number of new cases of disease by the sum of all disease-free time. Note that different participants contribute different amounts of time to the denominator. **EXAMPLE 1-5** is a purposely small example that illustrates the computations.

EXAMPLE 1-5

Suppose we wish to estimate the 10-year incidence of CVD among adults aged 45–55 who have a family history of CVD. We enroll eight participants between the ages of 45 and 55 years who have a family history of CVD and who themselves are free of CVD at the time of enrollment into the study. Each participant agrees to be followed for up to 10 years for the development of CVD. **TABLE 1-5** summarizes the experiences of each participant in the study, and our goal is to use the data in Table 1-5 to estimate the 10-year incidence of CVD.

TABLE 1-5 Experiences of Eight Patients over a 10-Year Follow-up Period

Participant 1	Develops CVD 6 years into the study
Participant 2	Drops out of the study after 8 years, at which time they are free of CVD
Participant 3	Completes the 10-year follow-up disease free
Participant 4	Develops CVD 2 years into the study
Participant 5	Dies of an injury 7 years into the study, at which time they are free of CVD
Participant 6	Completes the 10-year follow-up disease-free
Participant 7	Develops CVD 9 years into the study
Participant 8	Drops out of the study after 2 years, at which time they are free of CVD

Note that if we had 10 years of complete follow-up data on each participant (specifically if we were able to follow participants 2, 5, and 8 for the full 10 years and they were all disease-free after 10 years), then we could estimate the 10-year incidence proportion, or risk, of CVD as $3/8 = 0.375 = 37.5\%$. Unfortunately, we do not know the 10-year outcome status for participants 2, 5, and 8 and cannot assume that they were disease-free at 10 years. Instead, we compute a 10-year incidence rate by dividing the number of new cases of disease by the sum of all disease-free time. The 10-year incidence rate is computed as follows: $3/(6 + 8 + 10 + 2 + 7 + 10 + 9 + 2) = 3/54 = 0.056$ new cases per person year. This is equivalent to reporting 56 new cases of CVD per 1,000 person years.[d]

---

d Person years is a unit of measurement that takes into account the number of people in a study and also the amount of time each is assessed during the study (e.g., a study that

The incidence rate has an inherent time component (the denominator here is person years) compared to prevalence and incidence, which are proportions. Many people use the term "rate" when they are actually reporting proportions (e.g., prevalence rate). A rate describes the speed at which something happens (e.g., specifically transitioning from a disease free to a diseased state) and is specific to a particular time frame.

Ratios

Ratios describe the relative magnitude of two different measures and are computed by dividing one quantity by another. The numerator and denominator are not necessarily related, as is the case with a proportion where the numerator is a subset of the denominator. In Example 1-5 we estimated the 10-year incidence rate of CVD among adults aged 45–55 years with a family history of CVD as 56 new cases of CVD per 1,000 person years. Suppose we replicate the study among persons who do not have a family history of CVD and estimate the 10-year incidence rate of CVD among adults 45–55 years of age who do not have a family history of CVD as 21 new cases of CVD per 1,000 person years. The ratio of these incidence rates is computed as follows: 56 cases per 1,000 person years among adults 45–55 years of age with a family history of CVD to 21 cases per 1,000 person years among adults 45–55 years of age without a family history of CVD = 0.056/0.021 = 2.67. The 10-year incidence rate of CVD among adults between the ages of 45 and 55 with a family history of CVD is 2.67 times that of adults between the ages of 45 and 55 without a history of CVD. The ratio is a relative measure and interpreted as the resultant measure to one (e.g., 2.67 to 1, which can also be written as 2.67:1).

EXAMPLE 1-6 illustrates the computations of risks, rates, and ratios and their interpretations.

EXAMPLE 1-6

The Centers for Disease Control and Prevention report prevalence and incidence of diabetes mellitus in the United States in their 2017 diabetes report card.[5] **FIGURE 1-1** illustrates prevalence and incidence of diabetes among adults in the United States over the past 35 years. Note that the estimates are age-adjusted, which means that they are based on a standard population, specifically the U.S. population in 2000, which allows for fair comparisons over time.

(continues)

follows 100 people for 1 year each includes 100 person years of data as does a study that follows 10 people for 10 years each).

EXAMPLE 1-6   (*continued*)

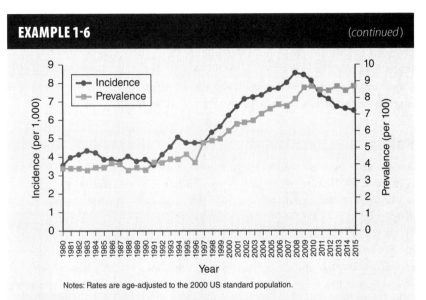

Notes: Rates are age-adjusted to the 2000 US standard population.

FIGURE 1-1 Prevalence and incidence of diabetes in adults from 1980 to 2015 in the United States.

Centers for Disease Control and Prevention, United States Diabetes Surveillance System and National Health Interview Survey, https://www
.cdc.gov/diabetes/library/reports/reportcard/incidence-2017.html

Incidence of diabetes mellitus increased steadily from a low of about 3.5% in 1991 to a high of over 8% in 2008, and then started to decline to about 6.5% in 2015. Meanwhile, the prevalence of diabetes mellitus has been increasing over time, leveling off at about 9% in 2015. In 2015, approximately 30.3 million people in the United States had diabetes mellitus.

Prevalence of diabetes mellitus varies by race/ethnicity, as seen in **FIGURE 1-2**. Again, age-adjusted estimates account for different age structures within each race/ethnicity group, allowing for fairer comparisons by race/ethnicity.

Using the data in Figure 1-2, we can compare the prevalence of diabetes mellitus in American Indian/Alaskan Natives to prevalence among non-Hispanic whites using a ratio: 0.151/0.074 = 2.04. Thus, the prevalence of diabetes among American Indian/Alaskan Natives is twice that in non-Hispanic whites in 2013–2015.

The same CDC report also summarizes annual incidence rates of diabetes mellitus among youth, aged 10–19 years, in 2011–2012 (**FIGURE 1-3**). The highest incidence rate is among American Indians, with 46.5 new cases per 100,000 in 2011–2012. The incidence rate among black youth is 32.6 new cases per 100,000 in 2011–2012 compared to 3.9 new cases per 100,000 in 2011–2012 among whites. It is perfectly appropriate to present these absolute incidence rates. Ratio estimates are relative comparisons and are often used to highlight inequities. For example, the incidence rate of diabetes mellitus among black youth is 8-fold higher than the incidence rate among whites (32.6/3.9 = 8.4). The latter might be a more compelling way to present these data to initiate additional inquiry and action to address why this might be the case.

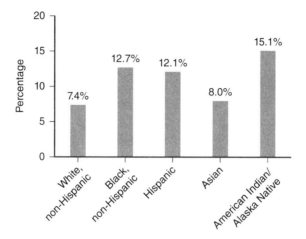

Notes: Percentages are age-adjusted to the 2000 US standard population.
Figure adapted from the *National diabetes statistics report, 2017.*

FIGURE 1-2 Age-adjusted prevalence of diabetes in the United States in 2015 by race/ethnicity.

National Diabetes Statistics Report, 2017. Data Sources: 2013–2015 National Health Interview Survey and 2015 Indian Health Service National Data Warehouse (American Indian/Alaska Native data), Centers for Disease Control and Prevention.

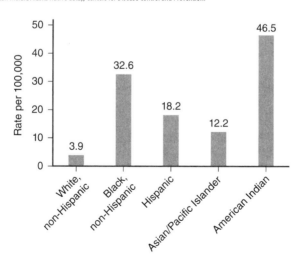

Note: American Indian youth who participated in the SEARCH study are not representative off all American Indian youth in the United States. Thus, these rates cannot be generalized to all American Indian youth nationwide.

FIGURE 1-3 Incidence of diabetes in children 10–19 years of age in the United States in 2011–2012.

SEARCH for Diabetes in Youth Study, Centers for Disease Control and Prevention, https://www.cdc.gov/diabetes/library/reports/reportcard/diabetes-in-youth-2017.html

Reporting risks, rates, or ratios depends on the question of interest and the data available to address that question. It is always critical to consider the audience. What data do they need? What information will be useful for action? And, as always, we must be precise in our terminology and how we present the data.

▶ 1.4 Graphical Displays of Data

Graphical displays are powerful tools to convey data and statistical information. To be effective, however, graphical displays must be clear, accurate, and honest. There are many tools and programs for creating sophisticated data visualizations (e.g., Tableau), but many highly effective data visualizations are carefully crafted variations of a few simple chart types available in many widely available platforms (e.g., Excel).

We propose four steps in producing and disseminating high-quality graphical displays. Step one is understanding your audience. Are they health professionals, policy makers, or community members? To understand your audience, ask: Who are they? What is their understanding of statistical/ technical issues (as this informs how technical a graphical display might be)? What data do they need to make decisions or to engage others to take action? How do they need the information? What do they already know on the topic?

Step two is creating a story. When presenting data or statistical results, it is important to provide background and context. Frame the problem, describe your approach to addressing it, and provide sufficient detail to both interest and engage your readers in the work.

Step three is designing your graphical display. This is informed by steps one and two.

Step four is disseminating your results. This might be in a peer-reviewed publication, a presentation at a professional meeting, a report for professionals in your or another field, or a presentation to community members at a local gathering. The way in which results will be disseminated also informs the design of the graphical display, which we will discuss in more detail.

When designing graphical displays it is important to choose the right type of display for the data and statistical results, use clear purposeful titles and labels to highlight key trends or associations, be honest with the data by always following sound statistical practices, and to generate interest in the topic. Edward Tufte discusses two important issues that should be considered in creating graphical displays—the data–ink ratio and "chart junk."[6] Tufte argues that designers should maximize ink dedicated to data as opposed to other design features as the data are the most important element in the display. Tufte also argues that designers should minimize "chart junk"—extraneous elements that distract the reader from the data.

And, just as was the case with summary statistics, the most effective graphical display depends on the variable type and the purpose of the visualization. For the latter, there are a few common uses for data visualizations: displaying distributions (i.e., responses to a particular measurement in a sample or group), comparing groups in terms of a particular risk factor or outcome, summarizing associations between risk factors and outcomes, and illustrating trends over time.

We use data in the study described in **EXAMPLE 1-7** to illustrate different graphical displays.

EXAMPLE 1-7

A study is conducted to evaluate risk factors for premature birth (i.e., birth of a baby before 37 weeks' gestation). In the United States, approximately 11% of babies are born prematurely. There are a number of factors that increase risk for premature birth, including prior premature birth, short duration between pregnancies, multiple gestations (e.g., twins, triplets), and high blood pressure.[7] A total of $n = 486$ women with prior premature birth histories are enrolled in the study during their first trimester of pregnancy. In this study, demographic data are collected at enrollment, and each woman is followed through the outcome of her pregnancy. We use data from this study to generate graphical displays for different variable types and purposes.

Distributions

This type of graphical display is used to convey the range of responses to a particular measurement or attribute observed in a sample. The distribution of a dichotomous variable is often not displayed graphically but reported in text form or summarized in a table. For example, in the study described in Example 1-7, each woman's age in years is recorded at enrollment. Maternal age is a continuous variable, but women 35 years of age and older are often considered at higher risk for various complications during pregnancy.[8] In the study, a dichotomous variable is created to classify study participants as <35 years versus 35 years and older. In the study, $398/486 = 82\%$ are less than 35 years of age and $88/486 = 18\%$ are 35 years of age and older. (Recall that this is a high-risk sample of women with a history of premature birth.) **FIGURE 1-4** displays the distribution of this dichotomous variable. A graphical display is probably not necessary to convey this information, but note some of the design features in Figure 1-4 that facilitate interpretation. The use of direct labelling (i.e., adding the percentages at the top of each bar) and the axis labels clearly detail what is being summarized.

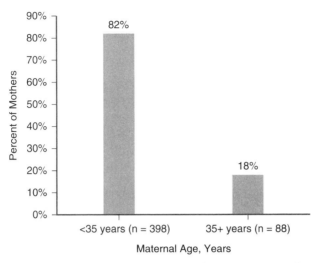

FIGURE 1-4 Percentage of mothers with a history of premature birth who are 35 years of age and older.

In the study, investigators also recorded race/ethnicity and all women self-identified as black, Hispanic, or white. **FIGURE 1-5** displays the distribution of race/ethnicity (a categorical variable) using a bar chart (note the spaces in between bars, as opposed to bars running together, which is a feature of a histogram used for ordinal variables). Again we use direct labelling and axis labels to clearly detail what is being summarized.

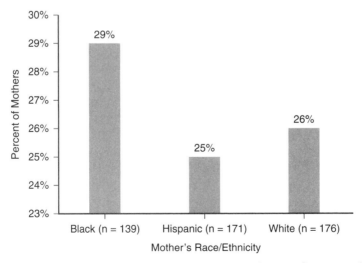

FIGURE 1-5 Distribution of race/ethnicity in mothers with a history of premature birth.

The primary outcome of the study is recurrent premature birth. Another key outcome is infant birthweight measured in grams. Birthweight is a continuous variable, but ordinal categories are often used to summarize birthweight as follows: extremely low birthweight <1,000 g (2 pounds, 3 ounces), very low birthweight <1,500 g (3 pounds, 5 ounces), low birthweight <2,500 g (5 pounds, 8 ounces), and normal birthweight 2,500–4,000 g (5 pounds, 8 ounces to 9 pounds, 4 ounces). **FIGURE 1-6** displays the distribution of birthweight categories (an ordinal variable) using a histogram (note that there are no spaces in between bars as adjacent ordinal categories reflect an underlying continuum, as opposed to a bar chart that has spaces in between bars as the response categories are unordered). Again we use direct labelling and axis labels to detail what is being summarized. We use the footnote at the bottom of the graphical display to define the different birthweight categories as we want to create visualizations that stand alone. Many readers often turn to tables and figures in papers and reports before reading the entire text; thus, it is important that tables and figures can stand alone and do not require the reader to search through the text to find details or definitions such as the definitions of the birthweight categories.

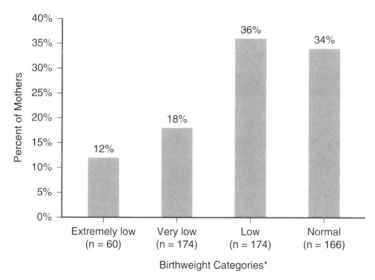

*Note: Extremely low birthweight <1,000 grams (2 pounds, 3 ounces), very low birthweight <1,500 grams (3 pounds, 5 ounces), low birthweight <2,500 grams (5 pounds, 8 ounces), and normal birthweight 2,500 to 4,000 grams (5 pounds, 8 ounces to 9 pounds, 4 ounces).

FIGURE 1-6 Distribution of birthweight categories in mothers with a history of premature birth.

As noted previously, maternal age in years was measured at enrollment. **FIGURE 1-7** displays the distribution of maternal age (a continuous variable) using a box and whisker plot (sometimes called a box plot). Box plots can be oriented vertically (as in Figure 1-7) or horizontally

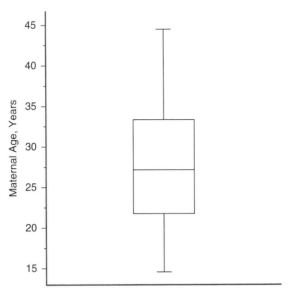

FIGURE 1-7 Distribution of mothers' ages, in years, in mothers with a history of premature birth.

(as in **FIGURE 1-8**). In Figure 1-7, the lower horizontal line is the minimum maternal age (15 years), and the upper horizontal line is the maximum maternal age (45 years). The box is framed by the lower or first quartile (22 years) and the upper or third quartile (33 years). The line across the middle of the box is the median maternal age (27 years). The box contains the middle 50% of the mothers' ages in this sample.

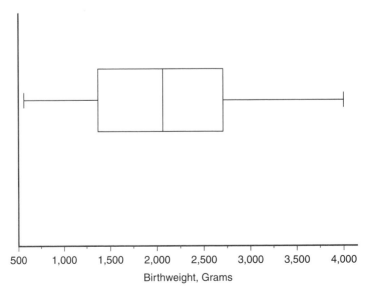

FIGURE 1-8 Distribution of birthweights, in grams, of babies born to mothers with a history of premature birth.

Figure 1-8 is an example of a horizontal box plot of birthweights, measured in grams. In the study sample, birthweights range from 600 to 4,000 g. The median birthweight is 2,065 g and the IQR (first quartile to third quartile) is 1,377–2,710 g. The choice of horizontal versus vertical box plot is a matter of preference. It is always good practice to try a few different formats and get feedback from colleagues and others to ensure that the ultimate design is one that most clearly and accurately summarizes the data.

Comparing Groups

These graphical displays are extremely valuable in comparing distributions of risk factors or outcomes among important subgroups. Again, depending on the nature of the variables (dichotomous, categorical, ordinal, continuous) being compared, different displays are appropriate. Continuing with Example 1-7, suppose we want to compare the percentages of mothers 35 years of age and older by race/ethnicity. **FIGURE 1-9** displays the percentages of mothers in each race/ethnicity group who are 35 years of age and older. And while Figure 1-9 makes it apparent that a higher percentage of Hispanic mothers are 35 years of age and older compared to mothers of other racial/ethnic backgrounds, these data might be better reported in a table as it may not be the best use of a graphical display, violating our data–ink principle (too much ink for very little data!).

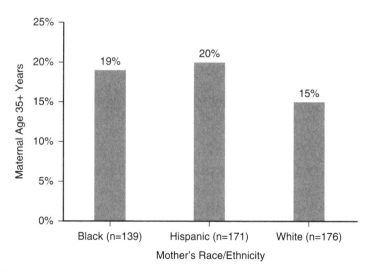

FIGURE 1-9 Percentages of mothers with a history of premature birth who are 35 years of age and older by race/ethnicity.

FIGURE 1-10 displays the distribution of birthweight categories by the mother's race/ethnicity. Note that within each racial/ethnic group, the percentages add up to 100%. From the figure it is clear that the distributions of birthweight categories vary by race/ethnicity, with babies born to

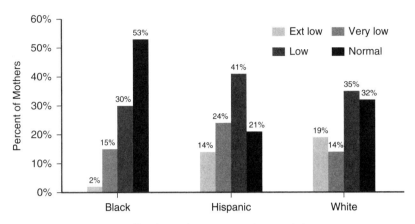

FIGURE 1-10 Distribution of birthweight categories by race/ethnicity among mothers with a history of premature birth.

Hispanic and white mothers in this high-risk sample being more likely to have extremely low birthweights (14% and 19%, respectively), compared to babies born to black mothers (2% extremely low birthweight).

For comparison purposes, consider **FIGURE 1-11,** which is another representation of the same data displayed in Figure 1-10. Note in Figure 1-11 that the bars do not run together (as in Figure 1-10, which shows side-by-side histograms of birthweight categories) as it displays the proportions of mothers within each racial/ethnic group in each birthweight category. Which is the better display of the data? As a reader/consumer of this data, which display is more useful?

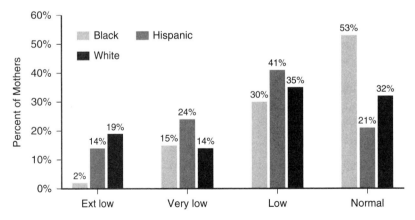

FIGURE 1-11 Distribution of birthweight categories by race/ethnicity among mothers with a history of premature birth—alternate display.

And consider yet another option to compare birthweight categories by mothers' race/ethnicity in **FIGURE 1-12**. Figure 1-12 shows the distribution of mothers' race/ethnicity within each birthweight category. The data displayed in Figure 1-12 are not the same as the data displayed in Figures 1-10

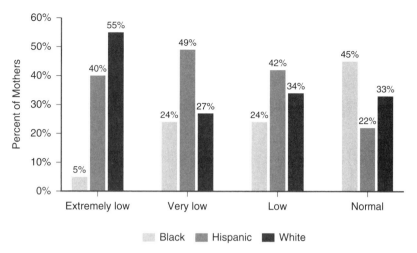

FIGURE 1-12 Distribution of race/ethnicity by birthweight categories among mothers with a history of premature birth.

and 1-11. Note in Figure 1-12 that the percentages of racial/ethnic groups add up to 100% within each birthweight category.

Which of these options is the best display of these data? The answer depends on the question being asked. If the question is about the distribution of birthweight categories within each racial/ethnic group, then either Figure 1-10 or Figure 1-11 is appropriate. If the question is about the racial/ethnic composition of each birthweight category, then Figure 1-12 is the best option. When preparing graphical displays, we aim to always make it easy for the reader to make the relevant comparisons.

The data to produce Figures 1-10 through 1-12 are often organized in a cross-tabulation table. Cross-tabulation tables are used to summarize the association between two dichotomous, categorical, or ordinal variables. **TABLE 1-6** is the cross-tabulation of race/ethnicity and birthweight category

TABLE 1-6 Cross-Tabulation of Race/Ethnicity and Birthweight Categories: Frequencies

	Birthweight Categories				
Frequency	Extremely Low	Very Low	Low	Normal	Total
Black	3	21	41	74	139
Hispanic	24	42	73	37	176
White	33	23	60	55	171
Total	60	86	174	166	486

using the data from the study described in Example 1-7. Each cell of the table contains the frequency or number of women in that race/ethnicity and birthweight category combination (note the legend in the top left corner of the table indicating that frequencies are displayed).

TABLE 1-7 is the same table but includes three additional numbers in each cell of the table. The top number in each cell is the count (or frequency). The second number in each cell is the overall percentage in each race/ethnicity and birthweight category. Note that the sum of these 12 percentages is 100% and, for example, 0.6% of the babies in this sample were born to black mothers and had extremely low birthweight (3/486 = 0.006 = 0.6%). The third number in each cell is the row percentage, which is computed by dividing the cell frequency by the row total shown in the far right column. Note that the row percentages add up to 100% in each row of the table. For example, 15.1% of the babies born to black mothers had very low birthweight (21/139 = 0.151 = 15.1%). The fourth number in each cell is the column percentage, which is computed by dividing the cell frequency by the column total shown in the bottom row. Note that the column percentages add up to 100% in each column of the table. For example, 44.6% of the normal weight babies were born to black mothers (74/166 = 0.446 = 44.6%). The decision to report or to display graphically the overall, row, or column percentages depends on the question of interest. Note the legend in the top left corner of the table that describes each cell entry. This type of cross-tabulation table is produced directly in many statistical computing packages. Note that percentages from Table 1-7 are displayed in Figures 1-11 and 1-12.

TABLE 1-7 Cross-Tabulation of Race/Ethnicity and Birthweight Categories: Frequencies, Overall, Row, and Column Percentages

	Birthweight Categories				
Frequency Overall % Row % Column %	Extremely Low	Very Low	Low	Normal	Total
Black	3	21	41	74	139
	0.6	4.3	8.4	15.2	28.6
	2.2	15.1	29.5	53.2	
	5.0	24.4	23.6	44.6	
Hispanic	24	42	73	37	176
	4.9	8.6	15.0	7.6	36.2
	13.6	23.9	41.5	21.0	
	40.0	48.8	42.0	22.3	

(continues)

TABLE 1-7 Cross-Tabulation of Race/Ethnicity and Birthweight Categories: Frequencies, Overall, Row, and Column Percentages   (*continued*)

White	33	23	60	55	171
	6.8	4.7	12.4	11.3	35.2
	19.3	13.5	35.1	32.2	
	55.0	26.7	34.5	33.1	
Total	60	86	174	166	486
	12.4	17.7	35.8	34.2	

FIGURE 1-13 displays the distributions of birthweights (a continuous variable), in grams, by race/ethnicity. Side-by-side box plots, oriented either horizontally or vertically (as shown in Figure 1-13), are an effective way to compare distributions of a continuous variable among groups. Note that the ranges of birthweights (smallest to largest) are similar across racial/ethnic groups, whereas the medians are quite different. Babies born to black mothers have a median birthweight of 2,520 g, whereas babies born to Hispanic and white mothers have median birthweights of 1,777 g and 1,920 g, respectively.

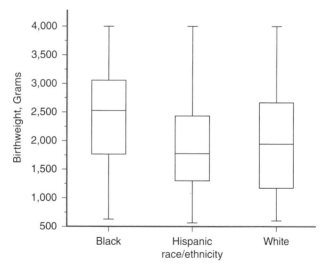

FIGURE 1-13 Side-by-side box plots of birthweights, in grams, by race/ethnicity in mothers with a history of premature birth.

Summarizing Associations

Associations between risk factors and outcomes are often summarized in graphical displays. Once again, depending on the nature of the variables (dichotomous, categorical, ordinal, continuous), different displays are appropriate. Continuing with Example 1-7, suppose we wish to summarize the association between maternal age (<35 vs. 35+ years) and birthweight, in grams. The mean birthweight of babies born to mothers <35 years of age is 2,082 g with a standard error (SE) of 46 g compared to a mean of 2,025 g with a standard error of 87 g among babies born to mothers 35 years of age and older. **FIGURE 1-14** displays the mean birthweights in each age group along with standard error bars. Recall that standard errors quantify the variability in summary statistics, such as the sample mean. Note that the *y*-axis in Figure 1-14 is scaled from 0 to 2,500 g, which is honest but may mask differences in birthweights that are clinically important. While it is critical to never intentionally mislead a reader by rescaling the *y*-axis, a rescaled *y*-axis might be necessary to better highlight differences. And importantly, if the *y*-axis is rescaled, this should be clearly noted on the *y*-axis (see Figure 1-15). There is another important issue with Figure 1-14 that relates to the data–ink ratio. It is generally not appropriate to display mean values using bars. The mean is one point; therefore, with a bar we are using far too much ink.

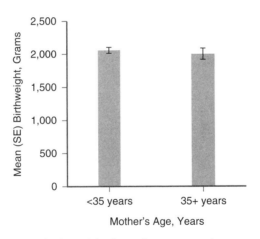

FIGURE 1-14 Mean (standard error) birthweights, in grams, by age among mothers with a history of premature birth.

FIGURE 1-15 is a better display of the mean birthweights and standard errors of the birthweights by age group using dots to indicate the mean values and a rescaled *y*-axis to better visualize the data. Note also the hash marks (//) on the *y*-axis to draw the reader's attention to the fact that the *y*-axis has been rescaled.

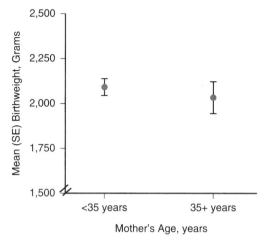

FIGURE 1-15 Mean (standard error) birthweights, in grams, by age among mothers with a history of premature birth—version 2.

FIGURE 1-16 is a scatter plot, which is a popular graphical display used to illustrate the association between two continuous variables. Each point in the scatter plot represents one participant's data. Figure 1-16 displays the association between gestational age (in weeks) and birthweight (in grams) and suggests a strong positive association between gestational age and birthweight. The convention in creating scatter plots is to show the risk factor on the x-axis and the outcome on the y-axis.

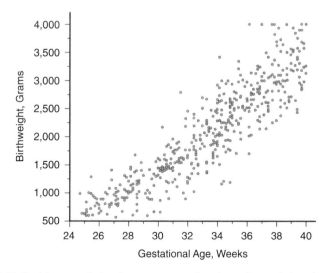

FIGURE 1-16 Positive association between gestational age (in weeks) and birthweight (in grams), among mothers with a history of premature birth.

FIGURE 1-17 is a scatter plot showing the lack of association between maternal age (in years) and birthweight (in grams) in this study sample.

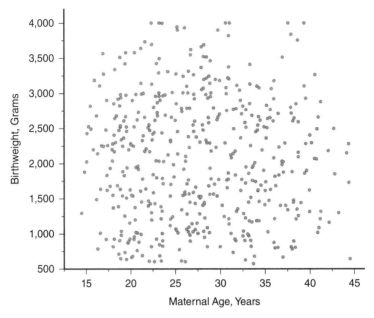

FIGURE 1-17 No association between maternal age (in years) and birthweight (in grams) among mothers with a history of premature birth.

Trends over Time

Trends over time are readily displayed graphically. Again, the exact statistics displayed over time depend on the type of variable being analyzed. For example, with a continuous variable (e.g., birthweight, in grams) means (or medians) would be displayed; with a dichotomous variable (e.g., an indicator of low birthweight), proportions or percentages would be displayed. Along with summary statistics such as means and proportions, measures of variability such as standard errors should also be displayed so that readers understand the sampling variability in the estimates displayed (we will discuss this in more detail in Unit 2).

The National Center for Health Statistics compiles data on a number of outcomes, including births and deaths to "guide actions and policies to improve the health of the American people." **FIGURE 1-18** shows the percentages of singleton births in the United States from 2006 to 2016 that were low birthweight (i.e., birthweights less than 2500 g or 5 pounds, 8 ounces).[9] Note that the percentage is relatively stable over the period 2006–2016.

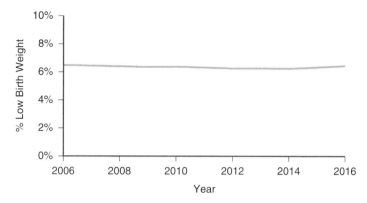

FIGURE 1-18 Percentages of low birthweights among singleton births in the United States, 2006–2016.

FIGURE 1-19 shows the same data, but with the *y*-axis rescaled from 6.2% to 6.5% (as opposed to 0%–10%) to highlight the differences over time. Again, when rescaling the *y*-axis, it is important to point this out to the reader so that data are not misinterpreted. With the rescaled *y*-axis it is clear to see that the percentage of low birthweight babies declined from 2006 to 2014 and then started to rise.

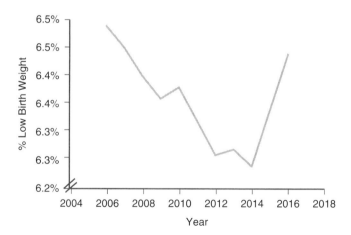

FIGURE 1-19 Percentages of low birthweights among singleton births in the United States, 2006–2016—version 2.

Note that the data displayed in Figures 1-18 and 1-19 are from the National Center for Health Statistics and are population-level data based on all births in the United States captured in the National Vital Statistics System. Thus, there are no estimates of sampling variability needed as these are not estimates from samples, but rather are population (or true) percentages. In contrast, **EXAMPLE 1-8** illustrates a scenario where estimates are based on samples and therefore the graphical display includes standard errors to quantify sampling variability.

EXAMPLE 1-8

A growth clinic at a local hospital enrolls $n = 42$ boys and $n = 28$ girls to participate in a study to monitor weights in children as they age from 5 to 10 years of age. Each child's weight is recorded annually, and the sample mean weights, in pounds, recorded at each age for boys and girls are shown in **FIGURE 1-20**. The error bars around each sample mean are standard errors computed by dividing the sample standard deviations by the respective sample sizes (i.e., $SE_{boys} = s/\sqrt{42}$ and $SE_{girls} = s/\sqrt{28}$, where s is the sample standard deviation in weight at each specific age).

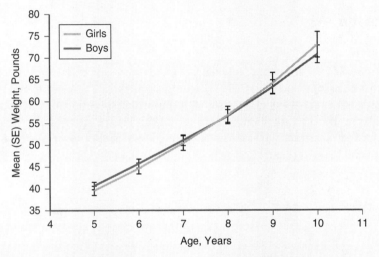

FIGURE 1-20 Mean (SE) weights in boys and girls from age 5 to 10 years.

▶ 1.5 Summary

Data, whether presented numerically or graphically, are extremely powerful and convincing when used appropriately. Understanding your audience is critical in preparing and reporting on data and statistical results. It is very important to use clear and correct terminology and to define important terms for those with less experience. It is also important to describe, discuss, and interpret data and statistical results in context.

We now return to the population health issue we started with in this unit—the opioid crisis in the United States—and we apply some of the techniques we discussed in this unit to address each question.

- ■ What are the characteristics of adolescents who misuse opioids?

  We start with the prevalence of opioid misuse in the United States. The U.S. Department of Health and Human Services Substance Abuse and Mental Health Services Administration (SAMSHA) collects, organizes, and makes available public use data on substance use, mental health, and a number of other issues. The 2017 National Survey on Drug Use and Health reports that opioid misuse varied by age group as follows: 3.1% of people aged 12–17 years, 7.3% of people aged 18–25 years, and 3.8% of people aged 26 years and older misused opioids.[10]

The U.S. Department of Health and Human Services Office of Adolescent Health reports that "all adolescents are at risk for misusing opioids," but that those with acute and chronic pain, mental illness, or using other substances such as alcohol or other drugs are at higher risk for opioid misuse. Protective factors (i.e., those that reduce the risk of opioid misuse) include strong family support, academic achievement, and an understanding of the risks and consequences of opioid and other drug use.[11]

The American Society of Addiction Medicine's 2016 fact sheet on opioids reports that in 2015, more than 275,000 adolescents were using pain relievers nonmedically, and more than 40% of these users were addicted to prescription pain relievers.[12]

- Do overdose deaths vary across states in the United States?

The Centers for Disease Control and Prevention has an interactive graphical display that allows a user to visualize drug overdose deaths and specific points in time and over time in each state in the United States.[13] **FIGURE 1-21** shows the variability in drug

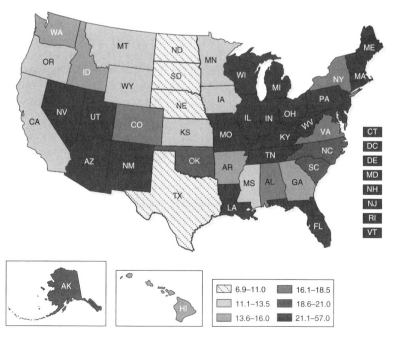

FIGURE 1-21 Variability in drug overdose deaths per 100,000 people in U.S. states in 2017.

overdose deaths (which are primarily driven by opioids) in the United States in 2017 using a choropleth map.[e]

■ What statistics and data visualizations might be effective to inform adolescents about the risks of opioid misuse?

To design an effective data visualization, we would need to understand what motivates adolescents—what information they would like to see that would be sufficiently convincing and relevant to prevent opioid misuse. There are many studies investigating this important issue. For example, a study by Miech et al. reports that what they call "legitimate" opioid use before graduation from high school is associated with a 33% increase in risk of opioid misuse after graduation.[14] Would this be an important finding to communicate to adolescents? If so, how? What other data would be important and how best would it be communicated?

As we conduct statistical analysis and consume statistical information, it is critical to understand context—who is involved in the study or analysis; specific attributes, characteristics, or conditions that might affect the results; and how the data should be interpreted. This relates to the generalizability of findings. We discussed the power of statistics, but statistical information is only relevant to the population from which the sample was drawn. Some people mistakenly overstate statistical results—they assume that findings in one particular subgroup imply that the same holds in others. This may, or may not, be the case. The issue of generalizability and appropriate interpretation of statistical results is a topic for further discussion and will be addressed in the next unit.

Key Points

■ In every statistical analysis it is important to classify variable types (dichotomous, categorical, ordinal, continuous) as statistical summaries, graphical displays, and statistical estimates and tests (which we discuss in the next unit) depend on variable types.

■ Mean values and medians represent typical values for continuous measurements. Medians are preferred when the measure is subject to extreme or outlying values.

■ Variability in a continuous measure is described by the sample standard deviation, s. The sample standard deviation quantifies variability around the sample mean (a larger standard deviation suggests more variability around the mean; a smaller standard deviation suggests that measurements are more tightly clustered around the sample mean).

■ Variability in a summary statistic (e.g., the sample mean) is described by the standard error. The standard error of the sample mean is

---

e A choropleth map uses different intensities of shading to display levels or quantification of another variable.

computed as $s/\sqrt{n}$ and quantifies variability in sample means, also called sampling variability.

- Prevalence is computed by dividing the number of existing cases (e.g., of a disease) by the total number of persons at risk. Prevalence is usually reported as a percentage and should be reported as prevalence at a specific point in time.

- Incidence, incidence proportion or risk, is computed by dividing the number of new cases (e.g., of a disease) by the total number of persons at risk. Incidence is usually reported as a percentage and should be reported as the incidence (or risk) of developing that disease over a specific time frame (e.g., annual incidence).

- The incidence rate is computed by dividing the number of new cases (e.g., of a disease) by the sum of all disease-free time. The incidence rate is often reported as the number of new cases per person years.

- Risks, rates, and ratios are specific measures that need careful attention to context and time.

- Ratios are relative measures that are useful in comparing prevalence, incidence, or incidence rates between unrelated groups (e.g., incidence of disease in blacks vs. whites, prevalence of a particular risk factor in people born in the United States vs. those born outside of the United States).

- Data visualizations can be extremely powerful, but must be designed with intention. They must be clear, be honest, adhere to sound statistical principles, and be able to stand alone.

References

1. National Institute on Drug Abuse. Opioid overdose crisis. https://www.drugabuse.gov/drugs-abuse/opioids/opioid-overdose-crisis. Accessed January 15, 2019.
2. Lohr SL. *Sampling: Design and Analysis.* 2nd ed. Boston, MA: Brooks/Cole Cengage Learning; 2010.
3. Rhee C, Lethbridge L, Richardson G, Dunbar M. Risk factors for infection, revision, death, blood transfusion and longer hospital stay 3 months and 1 year after primary total hip of knee arthroplasty. *Can J Surg.* 2018;61(3):165-176.
4. United States Census Bureau – American fact finder. https://factfinder.census.gov/faces/nav/jsf/pages/index.xhtml. Accessed December 15, 2018.
5. Centers for Disease Control and Prevention. Diabetes report card 2017. https://www.cdc.gov/diabetes/pdfs/library/diabetesreportcard2017-508.pdf. Accessed December 28, 2019.
6. Tufte ER. *The Visual Display of Quantitative Information.* 2nd ed. Cheshire, CT: Graphics Press; 2001.
7. Eunice Kennedy Shriver National Institutes of Child Health and Human Development. What are the risk factors for preterm labor and birth? https://www.nichd.nih.gov/health/topics/preterm/conditioninfo/who_risk. Accessed December 29, 2018.
8. March of Dimes. Pregnancy after age 35. https://www.marchofdimes.org/complications/pregnancy-after-age-35.aspx. Accessed March 2, 2019.

9. Centers for Disease Control and Prevention. Singleton low birthweight rates by race and Hispanic origin: United States, 2006–2016. https://www.cdc.gov/nchs/products/databriefs/db306.htm. Accessed January 2, 2019.

10. Substance Abuse and Mental Health Services Administration. Slides for the 2017 NSDUH annual national report. https://www.samhsa.gov/data/report/slides-2017-nsduh-annual-national-report. Accessed March 2, 2019.

11. U.S. Department of Health and Human Services Office of Adolescent Health. https://www.hhs.gov/ash/oah/adolescent-development/substance-use/drugs/opioids/index.html#risk. Accessed March 2, 2019.

12. American Society of Addiction Medicine. https://www.asam.org/docs/default-source/advocacy/opioid-addiction-disease-facts-figures.pdf. Accessed December 29, 2018.

13. Center for Disease Control and Prevention. Drug overdose deaths. https://www.cdc.gov/drugoverdose/data/statedeaths.html. Accessed January 5, 2019.

14. Miech R, Johnston L, O'Malley PM, Keyes KM, Heard K. Prescription opioids in adolescence and future opioid misuse. *Pediatrics.* 2015;136(5):1169-1177.

UNIT 2

Associations Between Two Variables

In biostatistical applications, we are often interested in associations between a risk factor and a health outcome in a population. For example, is there an association between participation in college athletics and sudden cardiac arrest? Are there adverse health effects when following a strict plant-based diet? Is ketamine effective in treating depression in adolescents? Are there long-term health risks associated with vaping? To address questions such as these, we investigate the association in a sample that is ideally representative of the population, and we make inferences about the association in the population (the true association) on the basis of what is observed in the sample. Making this leap from the sample to the population requires an understanding of probability that allows us to quantify uncertainty, which is a critical element of statistical inference.

In this unit, we review basic principles of probability and discuss an important application of probability related to evaluating the performance of screening tests that are extremely important to population health for early detection of disease. We then discuss techniques to quantify uncertainty in statistical applications as a critical component of confidence interval estimates and tests of hypothesis, the key tools of statistical inference.

A Population Health Issue—Physical Inactivity Among Children Worldwide Is Raising Concerns

The World Health Organization reports that four preventable individual behaviors—smoking, physical inactivity, unhealthy diet, and harmful alcohol use—account for most noncommunicable diseases worldwide.[1] In 2018, the Active Healthy Kids Global Alliance graded 49 countries and 75% of countries had failing physical activity grades attributable to increasing screen time, urbanization, and automation.[2] The authors of the report expressed concern for the future health and welfare of these inactive children. In its 2018 report card, the United States received a grade of D– on overall physical activity.[3] Less than 25% of children engaged in the recommended 60 minutes of moderate-to-vigorous activity per day, and there was an inverse association between age and physical activity, with 43% of children aged 6–11 years meeting the recommendations compared to 8% and 5% of 12–15-year-olds and 16–19-year-olds, respectively. There was also a difference by gender, with 28% of boys meeting the recommendations compared to 21% of girls, and a difference by race/ethnicity, with 25% of white children 6–17 years of age engaging in 60 minutes of physical activity per day, 27% of African American children, and 22% of Hispanic children meeting the recommendations.

In this unit, we discuss biostatistical techniques that allow us to answer questions such as the following:

- How likely are children worldwide to meet the recommended physical activity guidelines?
- What are the health benefits and consequences of physical inactivity?
- What interventions might be effective in increasing physical activity in children?

▶ 2.1 Concepts and Applications of Probability

Probability is very important in thinking about health. For example, your probability or risk of developing heart disease in the future might be high or low based on your family history, where you live, and how you live. Why is it important to know your risk? If you have a high probability of developing heart disease—or any other disease—there may be steps you can take now to lower that risk. Preventing disease from occurring is always a better route than treating disease once it has occurred.

To compute the probability that an outcome might occur, we divide the number of participants with the outcome of interest by the total number of participants in the population. Probabilities range from 0 to 1, inclusive. A probability of 0 means that an outcome is never going to occur, and a probability of 1 means it is certain to occur. Applications of greatest interest have probabilities between 0 and 1. We illustrate the computations of probabilities and their interpretations through examples, starting with **EXAMPLE 2-1**.

EXAMPLE 2-1

The Centers for Disease Control and Prevention (CDC) collects data on a number of health conditions and makes these data available for public use. **TABLE 2-1** contains data from a report on the numbers of children aged 8 years with, and without, autism spectrum disorder (ASD) in each of the 11 U.S. states that participated in the Autism and Developmental Disabilities Network in 2014.[4]

(continues)

EXAMPLE 2-1   (*continued*)

TABLE 2-1 Numbers of Children With and Without Autism Spectrum Disorder (ASD) in 11 U.S. States in 2014

State	Number Free of ASD	Number with ASD	Total Population
Arizona	24,603	349	24,952
Arkansas	39,470	522	39,992
Colorado	40,556	572	41,128
Georgia	50,292	869	51,161
Maryland	9,756	199	9,955
Minnesota	9,533	234	9,767
Missouri	24,977	356	25,333
New Jersey	31,971	964	32,935
North Carolina	29,756	527	30,283
Tennessee	24,553	387	24,940
Wisconsin	34,543	494	35,037
All states combined	320,010	5,473	325,483

Using the data in Table 2-1, we can estimate probabilities. For example, what is the probability that an 8-year-old child living in one of these 11 states had ASD in 2014? We can represent this question as: $P(\text{ASD})$ = number of ASD cases/total population size = 5,473/325,483 = 0.017, or 1.7%. This probability represents the overall prevalence of ASD in these 11 states. We could also ask, what is the probability that an 8-year-old child had ASD and lived in Georgia? $P(\text{ASD and lives in Georgia})$ = number of ASD cases in Georgia/ total population size = 869/325,483 = 0.003, or 0.3%. Both of these probabilities are unconditional probabilities—they are based on the total population size ($N = 325{,}483$), which appears in the denominator of each probability.

Using the data in Table 2-1, we can also address questions such as: What is the probability that an 8-year-old child living in Arkansas was diagnosed with ASD in 2014? This is an example of a conditional probability—the conditioning refers to the focus on a specific subgroup, that is, those residing in Arkansas. We compute the probability in a similar way to the approach described above except that we first restrict our attention to those living in Arkansas. $P(\text{Child living in Arkansas has ASD})$ = number of ASD cases in Arkansas/population of Arkansas = 522/39,992 = 0.013, or 1.3%. Similarly, what is the probability that an 8-year-old child living in Minnesota was diagnosed with ASD in 2014? P(Child living in Minnesota has ASD) = number of ASD cases in Minnesota/population of Minnesota = 234/9,767 = 0.024, or 2.4%. We could also use these data to compute a prevalence ratio comparing cases of ASD in Arkansas to those in Minnesota: 0.013/0.024 = 0.542. The prevalence of ASD in Arkansas is about half that in Minnesota. What could explain this?

EXAMPLE 2-2 is another example that illustrates the calculation of conditional probabilities.

EXAMPLE 2-2

A population study is conducted to evaluate depression in 132,525 adults 50 years of age and older living in Boston in 2018. Each participant undergoes an extensive examination and is classified as depressed if they report major or severe depressive symptoms over the past 2 weeks or not. In addition, each participant is also asked whether either of their parents was ever diagnosed with depression. The results are summarized in **TABLE 2-2**.

TABLE 2-2 Depression in Individuals 50 Years of Age and Older Living in Boston in 2018 and Their Parents

	Depression	No Depression	Total
Parents diagnosed with depression	1,053	4,090	5,143
Parents free of depression	8,223	119,159	127,382
Total	9,276	123,249	132,525

What is the probability, or prevalence, of depression in adults 50 years of age and older living in Boston in 2018? P(depression) = 9,276/132,525 = 0.070 = 7.0%. What is the prevalence of depression among those with a parent diagnosed with depression? P(depression among those with a parent diagnosed with depression) = 1,053/5,143 = 0.205 = 20.5%. What is the prevalence of depression among those whose parents were never diagnosed with depression? P(depression among those with neither parent diagnosed with depression) = 8,223/127,382 = 0.065 = 6.5%. The prevalence ratio of depression in adults 50 years of age and older with versus without a parent diagnosed with depression is 0.205/0.065 = 3.15. Adults 50 years of age and older living in Boston in 2018 with a parent diagnosed with depression have more than 3-fold higher prevalence of depression compared to those who do not have a parent diagnosed with depression.

▶ **2.2 Screening and Diagnostic Tests**

Screening tests are used in public health for early detection of risk factors or diseases in people who have yet to experience signs or symptoms of disease. Screening tests are often simple, noninvasive tests compared to diagnostic tests, which can be expensive, invasive, carry associated risks, and are used to establish the presence or absence of disease in people who have a positive screening test or who exhibit other signs and symptoms related to disease.

For example, screening tests are used extensively in the first and second trimesters of pregnancy to evaluate whether the developing fetus might be at risk for certain birth defects or chromosomal disorders such as Down syndrome. Women who screen positive are then referred for more invasive diagnostic tests, such as chorionic villus sampling or amniocentesis, to confirm or rule out a diagnosis.

Another example of screening involves regular assessment of blood pressure and cholesterol levels at annual physical examinations. Patients with elevated blood pressure or cholesterol levels, and those who report chest pain, shortness of breath, or pain in their legs or arms might undergo more invasive diagnostic tests including blood tests, electrocardiogram (ECG), echocardiogram, or cardiac catheterization to rule out, or confirm, suspected cardiovascular disease.

Performance Measures of Screening Tests

In order to decide whether to take a screening test or to recommend a screening test, we must first understand whether the screening test is accurate. There are a number of measures that are used to describe the performance of screening tests, in particular how the results of a screening test (positive, negative) agree with the results of a diagnostic test (disease, no disease).

Screening tests can be based on continuous or quantitative measurements (e.g., protein concentration in a blood sample or a measured total serum cholesterol level) or a binary or dichotomous classification (positive, negative). With continuous measurements, investigators often define a threshold of positivity such that if a person measures at, or above, that value they are considered test positive, and if they measure below that value they are considered test negative (the reverse is true if lower scores on the test are associated with increased risk of disease). We illustrate how quantitative measurements are used in screening, but first establish some notation and define key performance measures.

The diagnostic test is considered the gold standard and is the true measure of disease or outcome status. Each participant is classified as having the disease or not, having or not having the outcome of interest based on the diagnostic test. Assessments of the accuracy of screening tests are based on the relationship between the results of the screening and diagnostic tests using data that is often organized as shown in **TABLE 2-3**. The letters in Table 2-3 are placeholders for the numbers of patients that fall into each cell of the 2×2 table. The sample size (n), or number of patients tested, appears in the bottom right corner of the table.

TABLE 2-3 Data to Summarize Performance of Screening Tests

	Disease	No Disease	Total
Screen +	a	b	$a+b$
Screen −	c	d	$c+d$
Total	$a+c$	$b+d$	n

Four measures of performance are calculated to summarize the accuracy of screening tests, as defined in **TABLE 2-4** and using the notation outlined in Table 2-3, and all are conditional probabilities.

TABLE 2-4 Assessing the Performance of Screening Tests

Performance Measure	Definition	Calculation
Sensitivity, true positive fraction	Probability that a person with the disease tests positive	$a/(a+c)$
Specificity, true negative fraction	Probability that a disease-free person tests negative	$d/(b+d)$
False-positive fraction	Probability that a disease-free person tests positive	$b/(b+d)$
False-negative fraction	Probability that a person with the disease tests negative	$c/(a+c)$

Note the relationships among the measures: sensitivity and the false-negative fraction sum to 1 as do the specificity and the false-positive fraction. Thus, not all four measures need to be reported. For many screening tests, the sensitivity and false-positive fractions are reported, as the false-negative fraction and specificity can be computed by subtraction (e.g., specificity = 1− false-positive fraction).

Deciding whether a particular screening test is accurate requires considering the implications of each of these performance measures. A perfectly accurate screening test would have sensitivity = specificity = 1 and false-positive fraction = false-negative fraction = 0. Unfortunately, this is very unlikely.

A highly sensitive test is one that often comes back positive in people with disease. Would a sensitivity of 80% be sufficiently high? A sensitivity of 80% means that 80% of people with the disease test positive. It also means that the same test has a false-negative fraction of 20%. Specifically, 20% of people with the disease test negative. Is this a problem? It depends on the implications.

A highly specific test is one that often comes back negative in people free of disease. Would a specificity of 70% be sufficiently high? This means that 70% of disease-free people test negative. It also means that the same test has a false-positive fraction of 30%, meaning that 30% of disease-free people test positive. Is this a problem?

To determine the levels of sensitivity and specificity, false-negative and false-positive fractions that would be acceptable depend on the nature of the disease or health outcome and whether missing it (false negative) or telling a patient that they screen positive when in reality they do not have the disease (false positive) has serious implications psychologically, medically, financially, and so on. A false-negative result might incorrectly reassure a patient and a false-positive result might unnecessarily stress a patient.

Determining acceptable levels of accuracy for any screening test depends on the nature of what is being tested (e.g., HIV, color blindness) and the implications of errors in screening. There are also trade-offs as these four performance measures are related.

Consider again the scenario where the screening test produces a continuous or quantitative assessment, as opposed to a binary or dichotomous classification (i.e., screen positive vs. screen negative). Investigators often determine a threshold of positivity (or cut-off point) so as to produce certain performance characteristics of the screening test. **EXAMPLE 2-3** illustrates the approach.

EXAMPLE 2-3

Syphilis is a sexually transmitted, chronic infectious disease caused by *Treponema pallidum* that is usually transmitted by sexual contact or from mother to child during childbirth. The prevalence of syphilis has been increasing worldwide.[5] In the United States in 2017, over 30,000 cases of syphilis were reported.[6] Syphilis is diagnosed with laboratory tests that include both non-*Treponema pallidum* (non-TP) and *Treponema pallidum* (TP) tests.[7] Screening is often performed with a sequence of tests consisting first of a non-TP test, followed by a TP test (or vice versa). A recent study of 1,000 Korean men and women who underwent screening at annual physical examinations reported a sensitivity of 98% and specificity of 57% for a screening test called *Treponema pallidum* latex agglutination (TPLA) that produced a quantitative result.[8] The investigators summarized their results with a receiver operating characteristic (ROC) curve, which is a visual display of sensitivity on the *y*-axis and the false-positive fraction, equivalent to 1-specificity, on the *x*-axis (**FIGURE 2-1**). Each of the points on the ROC curve represents the false positive fraction (i.e., 1-specificity), sensitivity combination at each distinct value of the TPLA measure.

(continues)

EXAMPLE 2-3   *(continued)*

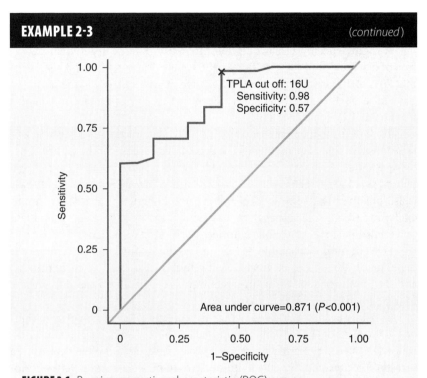

FIGURE 2-1 Receiver operating characteristic (ROC) curve.

Reproduced from Nah E-H, Cho S, Kim S, Cho H-I, Chai J-Y. Comparison of traditional and reverse syphilis screening algorithms in medical checkups. *Ann Lab Med.* 2017;37:511–515.

The authors determined that a threshold of positivity of 16 units (i.e., a TPLA value >16 U was considered test positive) produced the best performance of the screening test with a sensitivity of 98% and a specificity of 57%. Is this a good screening test? The sensitivity is high at 98%, but is the specificity of 57% sufficiently high? This means that this test has a false-positive fraction of 43% (i.e., 43% of patients free of syphilis screen positive). Is this acceptable?

EXAMPLE 2-4 provides another example of calculating and interpreting performance measures of a screening test.

EXAMPLE 2-4

Low-dose computerized tomography (CT) scans are used to screen for lung cancer and involve low doses of radiation to take detailed X-ray images of the lungs. The images are read by a radiologist and classified as needing further clinical examination or not. Diagnostic tests for lung cancer are based on sputum cytology, and are evaluated by a certified cytopathologist.

(continues)

EXAMPLE 2-4 *(continued)*

A sample of $n = 1200$ patients with a history of heavy smoking undergo both low-dose CT scan and sputum cytology. CT scans are read by radiologists and classified as needing further clinical examination (screen positive) or not (screen negative). An analysis of sputum cytology confirms the diagnosis of lung cancer, and the results of the screening and diagnostic tests are summarized in **TABLE 2-5**.

TABLE 2-5 Results of Low-Dose CT Scan and Sputum Cytology in $n = 1200$ Patients

	Lung Cancer	No Lung Cancer	Total
Screen +	29	93	122
Screen −	4	1,074	1,078
Total	33	1,167	1,200

The sensitivity of the test is $29/33 = 0.88$, or 88%. The specificity of the test is $1{,}074/1{,}167 = 0.92$, or 92%. The false-positive fraction is $93/1{,}167 = 0.08$, or 8%, and the false-negative fraction is $4/33 = 0.12$, or 12%.

Is low-dose CT scanning a good screening test for lung cancer? Although the sensitivity and specificity are high, it is often most useful to consider the implications of errors. This screening test has a false-positive fraction of 8%, meaning that 8% of patients who do not have lung cancer screen positive. Such news would clearly be stressful. All patients who screen positive would be referred for further testing—ideally this would happen quickly so that patients would learn their true (disease-free) state within a short period of time. This screening test also has a false-negative fraction of 12%, meaning that 12% of patients with lung cancer screen negative. The issue here is that these patients might be falsely reassured that they do not have lung cancer and would miss the opportunity for early diagnosis and treatment.

Unfortunately, there are no universal numerical thresholds to judge screening tests as good or not, or sufficiently accurate or not, as in some situations it is more important that the false-positive fraction is low (and the specificity high), while in others it is more important that the false-negative fraction is low (and the sensitivity high).

Understanding the sensitivity, specificity, false-positive fraction, and false-negative fraction of a screening test allows a patient to decide whether they want to take the screening test (i.e., whether they feel it will accurately detect disease if it is present or rule out disease if it is absent). There is another important aspect of screening that relates to how a patient inter-prets the result of a screening test once they have taken it. Two questions are of interest: (1) I took the screening test and it came back positive, should I worry that I have disease? (2) I took the screening test and it came back negative, should I be reassured that I am disease free?

TABLE 2-6 defines two additional performance measures that address these questions, again using the notation outlined in Table 2-3.

TABLE 2-6 Positive and Negative Predictive Values

Performance Measure	Definition	Calculation
Positive predictive value	Probability that a person who tests positive has the disease	$a/(a+b)$
Negative predictive value	Probability that a person who tests negative is free of the disease	$d/(c+d)$

Referring to Example 2-4, the positive predictive value of the CT scan is $29/122 = 0.24$, or 24%. Thus, the probability that a patient with a history of heavy smoking and a positive CT scan has lung cancer is 24%. Is this reason for concern? Based on the data in Table 2-5, the prevalence of lung cancer in patients with a history of heavy smoking is $33/1200 = 0.03 = 3\%$. This probability is called a pretest probability of lung cancer (i.e., the best estimate of the probability of lung cancer with no additional information) for this particular group. Once we have the results of the screening test—the CT scan in this case—we update our estimate to 24%, which is called a posttest probability. The knowledge of a positive screening test increases the patient's risk of having lung cancer 8-fold.

Positive and negative predictive values are highly influenced by the prevalence of disease. This notion often creates confusion for people interpreting screening tests. In Example 2-4, we determined that the CT scan for lung cancer had a sensitivity of 88% and a specificity of 92% (i.e., quite high). These measures describe the accuracy of CT scanning as a screening test. The likelihood that an individual will develop lung cancer is estimated at 3% (the observed prevalence in this group of individuals with a history of heavy smoking). Knowledge of a positive screening test substantially increases this risk.

The negative predictive value of CT scanning for lung cancer is $1,074/1,078 = 0.997$, nearly 100%. Thus, it is nearly certain that a patient with a negative CT scan does not have lung cancer. A negative CT scan provides strong reassurance.

When we have a complete cross tabulation on a specific issue (as in Examples 2-1 to 2-4), we can compute probabilities directly by dividing the number of cases that meet specific criteria by the total number in the relevant group. We sometimes modify the denominator if we are focusing on a specific subgroup (i.e., conditioning on a specific attribute), but we follow the same approach. There are other instances where the cross tabulation is not available, but we can use the knowledge about the performance of the screening test and the prevalence of disease to compute positive and negative predictive values.

Bayes' Rule

Bayes' rule is a probability rule often applied in practice whereby we update a prior probability of disease (e.g., prevalence of disease, also called pretest probability) by taking into account information that is available on the probability of certain signs and symptoms in people with and without disease (e.g., based on a screening test) to produce what is called a posterior (updated, or posttest) probability of disease. There are different versions of Bayes' rule, ranging from simple to complex. The simplest version of Bayes' rule is:

$$P(D+|S+) = \frac{P(S+|D+)\,P(D+)}{P(S+)},$$

where $D+$ represents the presence of disease ($D-$ represents absence of disease) and $S+$ represents a positive screening result ($S-$ represents a negative screening result). The vertical line ("|") in the probability statements indicates conditional probability—for example, $P(S+|D+)$ is the probability of a positive screening test among those with disease, or the sensitivity of the screening test.

A common application of Bayes' rule uses information on the overall prevalence of disease, $P(D+)$ and the likelihood or probability of certain signs and symptoms in people with the disease $[P(S+|D+)]$ and without the disease $[P(S+|D-)]$ to compute the probability that a person with a positive screening test or with specific signs and symptoms has the disease of interest $[P(D+|S+)]$. **EXAMPLE 2-5** illustrates this approach.

EXAMPLE 2-5

The prevalence of celiac disease, an autoimmune disease where patients have allergic reactions to gluten, is about 1 in 140 people in the United States.[9] There are several blood tests available to screen for celiac disease and these detect antibodies in the blood that react to gluten. A popular test is the tTG-IgA (tissue Transglutaminase immunoglobulin A) test, which comes back positive in approximately 2.6% of patients tested. Recent reports indicate that 93% of patients with celiac disease have positive tTG-IgA tests, compared to 2% of patients free of celiac disease who have positive tTG-IgA tests.[10] Given this information, what is the probability that a patient who screens positive on the tTG-IgA test has celiac disease?

The following information is available, organized using the notation we established here.

$P(D+)$ = Prevalence of celiac disease = $1/140 = 0.007 = 0.7\%$

$P(S+)$ = P(positive tTG-IgA test) = 2.6%

$P(S+|D+)$ = Sensitivity of tTG-IgA test = P(positive tTG-IgA test among patients with celiac disease) = 93%

$P(S+|D-)$ = False-positive fraction of tTG-IgA test = P(positive tTG-IgA test among patients free of celiac disease) = 2%

We wish to compute the positive predictive value = P(celiac disease among patients with positive tTG-IgA tests). Using Bayes' rule,

$$P(D+|S+) = \frac{P(S+|D+)\,P(D+)}{P(S+)} = \frac{0.93(0.007)}{0.026} = 0.25.$$

In this particular application, the prevalence of disease is the pretest probability of celiac disease (0.7%) and the positive predictive value is the posttest probability of celiac disease that is updated with the additional information from the screening test. Factoring in the information about the screening test results in a posttest probability of celiac disease of 25%, more than 35 times the pretest estimate.

Another approach, as an alternative to applying Bayes' rule, to compute the desired probability involves simulating a hypothetical population (when simulating, it is best to consider a large population) and then applying the known probabilities to produce a cross tabulation (see **TABLE 2-7**). The footnotes detail the steps applied in sequence to compute the cell entries.

TABLE 2-7 Hypothetical population simulation approach

	Celiac Disease	No Celiac Disease	Total
tTG-IgA test positive	65[4]	195	260[3]
tTG-IgA test negative	5	9,735	9,740
Total	70[2]	9,930	10,000[1]

Steps:

[1]Specify the (hypothetical) population size of 10,000.
[2]Compute the number of participants with celiac disease based on reported prevalence of 1 in 140 people = 0.007 (10,000 × 0.007 = 70).
[3]Compute the number of participants who test positive based on report that 2.6% of patients tested, test positive = 10,000 × 0.026 = 260.
[4]Compute the number of participants with celiac disease who test positive based on the test's sensitivity = 93% (70 × 0.93 = 65).
The remaining cell entries are computed by subtraction.

Once the cross-tabulation table is populated, we compute the probability that a patient who screens positive on the tTG-IgA test has celiac disease. P(celiac disease among patients with positive tTG-IgA tests) = 65/260 = 0.25.

▶ 2.3 Probability Models

There are many attributes about which we would like to make probability statements that are continuous or quantitative, as opposed to dichotomous, categorical or ordinal as per many of the previous examples. These continuous attributes follow different distributional forms. For example, **FIGURE 2-2** is an illustration of a uniform distribution where each value of a continuous measure, ranging from 1 to 20, is equally likely to occur.

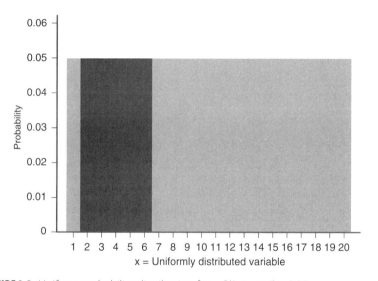

FIGURE 2-2 Uniform probability distribution for x, $P(2 \leq x \leq 6) = 0.25$.

With probability distributions, we compute probabilities using areas. Consider Figure 2-2 and suppose we ask, what is the probability of observing a value between 2 and 6? We express this as: $P(2 \leq x \leq 6)$ =? Here x represents the measurement of interest. Recall that the area of a rectangle is computed by multiplying base × height, thus the $P(2 \leq x \leq 6) = 5 \times 0.05 = 0.25$. (Note that the height of each bar is 0.05 as the total area sums to one.) There is a 25% probability of observing a value between 2 and 6, inclusive.

The Normal Distribution

The normal probability distribution is a popular distributional form that well represents many biological and health-related characteristics. The normal distribution (also called the "bell-shaped" distribution) has a peak in the middle and is symmetric around this center point (**FIGURE 2-3**). The normal distribution has several distinguishing features. First, the mean (note that we use μ to represent the population mean) = median = mode (the most frequently occurring value). Second, the distribution is symmetric about the mean (i.e., the distribution below the mean is a mirror image of the distribution above the mean). Third, about 68% of the area under the normal curve is between the mean plus 1 or minus 1 standard deviation (where σ = the population standard deviation[a]), about 95% of the area under the normal curve is between the mean plus 2 or minus 2 standard deviations, and almost all of the area under the curve is between the mean plus 3 or minus 3 standard deviations. Why is this important? If we know that a particular characteristic follows a normal distribution and we know its mean (μ) and standard deviation (σ), we actually know quite a bit about the distribution because these features hold for all normal distributions.

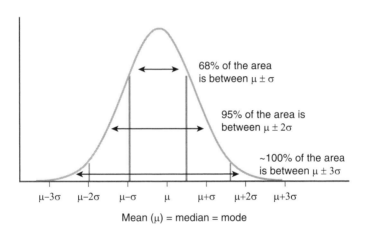

FIGURE 2-3 Properties of the normal distribution.

When adults undergo annual physical examinations, their primary care physicians often request a series of laboratory tests to assess blood counts, lipid levels, thyroid and liver functions, and so on. Laboratory test results are prepared for patients and often include notations to assist patients in interpreting their results. Because many of the assessments are approximately normally distributed for patients of the same sex and age, test results might include a note that indicates that the measured value was

a The population standard deviation is $\sigma = \sqrt{\dfrac{\Sigma(x-\mu)^2}{N}}$.

"within normal limits." What this means is that the measured value was within the mean plus 2 or minus 2 standard deviations, or among the middle 95% of all measured values for patients of the same sex and age. If an individual's result is outside of the normal limits, it may not be a problem as 5% of results would be expected in this range—it is simply a useful flag for ongoing monitoring.

The standard normal distribution can be used to compare values from different normal distributions (i.e., normal distributions with different means and standard deviations). **EXAMPLE 2-6** illustrates the approach.

EXAMPLE 2-6

The Framingham Heart Study is the world's longest running study of risk factors for cardiovascular disease.[11] In fact, many of the risk factors for cardiovascular disease were first discovered in the Framingham Heart Study. The study is supported by the National Heart, Lung, and Blood Institute, which also developed a publicly available teaching dataset from this landmark study. **FIGURES 2-4** to **2-7** illustrate the distributions of systolic blood pressure, diastolic blood pressure, total serum cholesterol, and body mass index in a sample of over 4000 participants in the teaching dataset based on the Framingham Heart Study.[12] In each figure, a normal distribution function has been superimposed. Note that each of these characteristics is approximately (i.e., not perfectly) normally distributed. Each distribution is subject to some extremes on the upper end (distributions with extreme values on the upper end are called right skewed), but not so many so as to shift the distributions too far away from the normal, or bell-shaped, distributional form.

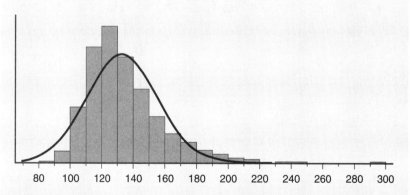

FIGURE 2-4 Systolic blood pressure: $\bar{X} = 132.9$ mmHg, $s = 22.4$ mmHg, range = 83.5–295.0 mmHg.

(continues)

EXAMPLE 2-6   (*continued*)

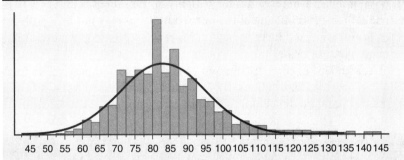

FIGURE 2-5 Diastolic blood pressure: $\bar{X} = 83.1$ mmHg, $s = 12.1$ mmHg, range = 48.0–143.0 mmHg.

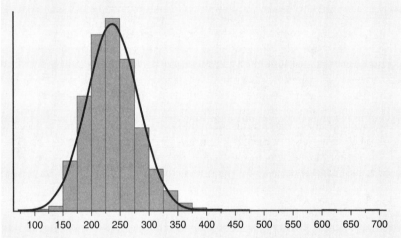

FIGURE 2-6 Total serum cholesterol: $\bar{X} = 237.0$ mg/dL, $s = 44.7$ mg/dL, range = 107.0–696.0 mg/dL.

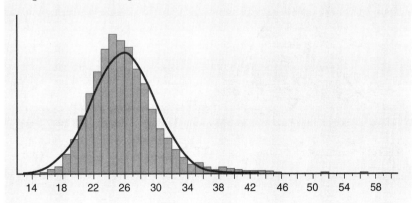

FIGURE 2-7 Body mass index: $\bar{X} = 25.8$ kg/m², $s = 4.1$ kg/m², range = 15.5–56.8 kg/m².

Suppose we focus on total serum cholesterol (Figure 2-6). There are some extremely high values in the sample (e.g., 696 mg/dL), but the distribution is well described by a normal distribution. In this large sample ($n = 4{,}382$), the mean total serum cholesterol is $\overline{X} = 237.0$ mg/dL and the median is 234 mg/dL. Note that the mean and median are close in value. In theory, if we had the full population and the distribution of total serum cholesterol followed a normal distribution, the mean and median would be identical. The standard deviation is 44.7 mg/dL. If we add and subtract two standard deviations (i.e., 89.4 mg/dL) from the mean ($237.0 - 89.4 = 147.6$ and $237.0 + 89.4 = 326.4$) we would expect 95% of the total serum cholesterol levels to fall within this range.

Consider the distribution of total serum cholesterol illustrated in Figure 2-6 with a mean of 237 mg/dL and a standard deviation of 44.7 mg/dL. Suppose we assume that this distribution mirrors the population with $\mu = 237$ mg/dL, $\sigma = 44.7$ mg/dL, and that total serum cholesterol follows a normal distribution. We wish to compute the proportions of participants with total serum cholesterol levels less than 200 mg/dL, between 200 and 240 mg/dL, and above 240 mg/dL, which are described by WebMD, a popular online resource for health information and advice, as "desirable," "borderline high," and "high" levels of total serum cholesterol, respectively. Specifically, we wish to compute the percentages of the Framingham Heart Study participants that fall into each of these three categories: $P(x < 200 \text{ mg/dL})$, $P(200 \le x \le 240 \text{ mg/dL})$, and $P(x > 240 \text{ mg/dL})$, where x represents total serum cholesterol level.

FIGURE 2-8 is the normal distribution of total serum cholesterol with $\mu = 237$ mg/dL, $\sigma = 44.7$ mg/dL, and depicts the thresholds, for the different total serum cholesterol categories.

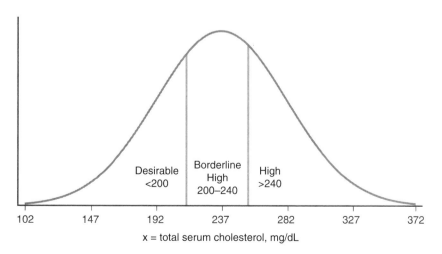

FIGURE 2-8 Categories of total serum cholesterol.

Probabilities for a normal distribution are computed as areas under the normal curve. In order to compute the desired probabilities, for example, $P(x < 200 \text{ mg/dL})$, we use what is called the standard normal distribution.

The Standard Normal Distribution

The standard normal distribution is a normal distribution with a mean (μ) of 0 and standard deviation (σ) of 1. We refer to the standard normal distribution with z so as not to confuse it with other attributes that follow a normal distribution that we generally reference with x. Approximately 68% of z scores are between -1 and 1, 95% are between -2 and 2, and almost all are between -3 and 3 (**FIGURE 2-9**).

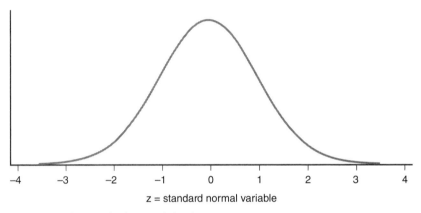

z = standard normal variable

FIGURE 2-9 The standard normal distribution, z.

The standard normal distribution is a specific normal distribution (with $\mu = 0$, $\sigma = 1$), and probabilities about the standard normal distribution are tabulated in many statistics and biostatistics textbooks, available in statistical computing packages and in Excel. We use the standard normal distribution to answer probability questions about other normal distributions. Specifically, we restate a question about any normal distribution, x, into a question about the standard normal distribution, z, and use tabulated values, a statistical computing or other package to answer the desired question.

Returning to Example 2-6, we are interested in $P(x < 200 \text{ mg/dL})$, where x represents total serum cholesterol level. This question can be restated as: $P\left(z < \dfrac{200 - 237}{44.7} = -0.83\right)$, where z represents the standard normal distribution and the conversion from x to z is: $z = \dfrac{x - \mu}{\sigma}$. This formula converts an x value from a normal distribution with mean μ and standard deviation σ into its corresponding z value. The z value or z score indicates the number of standard deviation units that the x value is from its mean, μ.

Once we have restated our question in terms of z, we can use published statistical tables, a statistical computing package or the norm.dist function in Excel[b] to compute the desired probabilities: $P(x < 200$ mg/dL$) = p(z < -0.83) = 0.204$. $P(200 \leq x \leq 240$ mg/dL$) = p(-0.83 \leq x \leq 0.07) = 0.323$ and $P(x > 240$ mg/dL$) = p(z > 0.07) = 0.473$. Approximately 20% of participants have desirable total serum cholesterol levels, 32% have borderline high and 47% have high total serum cholesterol levels (see Figure 2-8).

Using z Scores for Comparisons

The standard normal distribution can also be used to compare values from two different normal distributions. **EXAMPLE 2-7** illustrates the approach.

EXAMPLE 2-7

Heights for boys and girls at specific ages are approximately normally distributed (i.e., for specific sex and age groups, heights follow a bell-shaped curve). The CDC publishes height-for-age and weight-for-age data for boys and girls that are often used by pediatricians to monitor infant growth.[13] It is important to note that it is entirely possible for an infant to have a height-for-age outside of the mean plus 2 or minus 2 standard deviations; however, it is unlikely. In fact, approximately 5% of all infants will have height-for-age outside of the mean plus 2 or minus 2 standard deviations.

Focusing on boys, suppose that heights for 12-month-old boys are normally distributed with a mean height of 75.0 cm and a standard deviation of 2.5 cm and heights for 15-month-old boys follow a normal distribution with a mean height of 79.4 cm and a standard deviation of 2.9 cm.

A 12-month-old boy is 80 cm tall and a 15-month-old boy is 82 cm tall (**FIGURE 2-10**). Who is taller relative to his peers?

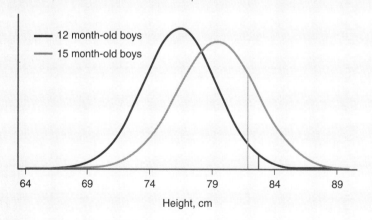

FIGURE 2-10 Comparing heights in 12- and 15-month-old boys.

---

b The norm.dist function requires the following arguments: norm.dist(x, mean, standard deviation, cumulative). Specifying cumulative $= 1$ returns the probability below x. For example, to compute $P(x < 200)$, we specify norm.dist(200,237,44.7,1). To compute $P(200 \leq x \leq 240)$, we specify norm.dist(240,237,4.7,1) − norm.dist(200,237,44.7,1).

In absolute terms, a height of 82 cm is taller than a height of 80 cm. However, the question is about relative heights. Is a height of 80 cm for a 12-month-old boy more extreme than a height of 82 cm for a 15-month-old boy? Figure 2-10 illustrates this comparison.

Comparing heights from two different normal distributions is like comparing apples to oranges. To make a fair comparison, we convert the heights (80 and 82 cm) into z scores and then compare the z scores directly.

We convert each height into a z score by subtracting the respective mean and dividing by the respective standard deviation. A height of 80 cm for a 12-month-old is 2.0 standard deviations above the mean height of 75.0 cm $\left(\dfrac{80 - 75.0}{2.5} = 2.0 \right)$, whereas a height of 82 cm for a 15-month-old is 0.9 standard deviations above the mean height of 79.4 cm $\left(\dfrac{82 - 79.4}{2.9} = 0.9 \right)$.

Thus, the 12-month-old boy with a height of 80 cm is taller relative to his peers than the 15-month-old boy with a height of 82 cm.

In the following sections, we combine what we have learned about descriptive statistics with these concepts of probability to move into statistical inference. Specifically, we describe two popular approaches to statistical inference—estimation and hypothesis testing. We describe each separately and then how they are related.

▶ 2.4 Estimation

In practice, we often wish to estimate an unknown population parameter (e.g., the mean of a continuous variable or measure or the proportion with a specific health outcome, a dichotomous variable) based on a single random sample from the population. We design studies so that the sample is as representative of the population as possible to make valid inferences about the population parameter based on sample statistics. Unfortunately, we never know just how representative our sample is because the population is not observed. Thus, in estimating population parameters, we must always account for uncertainty, specifically sampling variability.

The specific techniques to estimate unknown population parameters follow a general approach but are tailored specifically for each variable type (e.g., continuous vs. dichotomous outcomes). When we have a continuous outcome (e.g., total serum cholesterol level), we estimate the mean of that outcome in the population, μ. When we have a dichotomous outcome (e.g., diagnosis of diabetes), we estimate the proportion with the outcome of interest in the population, p. The sample mean $\left(\bar{X} = \dfrac{\Sigma x}{n} \right)$ is the best estimate of the true, unknown population mean (μ), and the sample proportion $\left(\hat{p} = \dfrac{\text{number with outcome}}{n} \right)$ is the best estimate of the true, unknown population proportion (p).

A good estimate of a population parameter is one that is close to the true value. If the measure we are estimating is highly variable (i.e., a continuous outcome with a large standard deviation), then it can be difficult to generate a precise estimate. If our estimate is based on a small sample size, our estimate may be imprecise. It is always important to generate estimates of true population parameters with as much precision as is possible or feasible. We always quantify and communicate the precision (or lack thereof) of any estimate so that consumers of the information can interpret it appropriately.

The Central Limit Theorem

An important theorem in statistics, the central limit theorem, supports the techniques we use in estimation and hypothesis testing. The central limit theorem states that when we have a large sample (usually $n \geq 30$ is sufficient), the distribution of all possible sample means of that size from a given population (regardless of the distribution of the outcome in the population) is approximately normally distributed. The same holds for sample means of any size if the outcome follows a normal distribution in the population. Sometimes we know this additional information about the population and thus can apply the results of the central limit theorem (described below) for samples of any size. If we cannot be sure that the outcome follows a normal distribution in the population, then we can use this result that the sample mean follows a normal distribution when $n \geq 30$. This is important because we can use what we know about probabilities of normal distributions in estimation and hypothesis testing.

Consider **FIGURE 2-11** which illustrates how the central limit theorem works using simulated data. In Figure 2-11, the distribution of the outcome, x, in the population ranges from 0 to 16 and does not follow a normal distribution. Rather, the distribution in the population is right skewed. The population mean, μ, is 3.0 and the population standard deviation, σ, is 1.7.

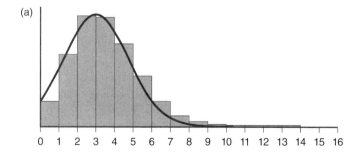

FIGURE 2-11 (Continues) (a) Distribution of outcome, x, in population.

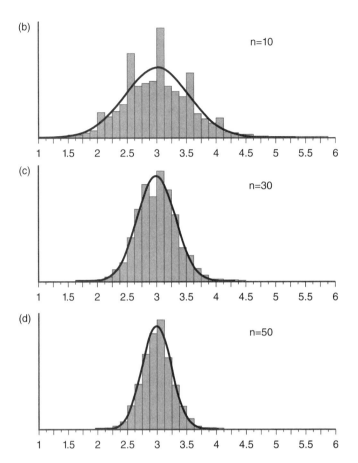

FIGURE 2-11 (Continued) (b) Distribution of sample means of size $n = 10$, (c) Distribution of sample means of size $n = 30$, and (d) Distribution of sample means of size $n = 50$.

If we take repeated samples of size 10 from the population, compute the sample mean, $\overline{X}$, for each sample, and plot those sample means, we get a differently shaped distribution. Sample means, based on samples of size $n = 10$, range from 1.0 to 5.8 and are not as right skewed. The mean of all sample means of size $n = 10$ is 3.0, and the standard deviation of the sample means, also called the standard error, is 0.5. (Recall that the standard error is computed as: $\dfrac{\sigma}{\sqrt{n}} = \dfrac{1.7}{\sqrt{10}} = 0.5$.)

If we take repeated samples of size 30 from the population, compute the sample mean, $\overline{X}$, for each sample, and plot those sample means, we again get a differently shaped distribution. Sample means, based on samples of size $n = 30$, range from 1.8 to 4.4 and look approximately normally distributed.

The mean of all sample means of size $n = 30$ is 3.0, and the standard deviation of the sample means, or the standard error, is 0.3 $\left(\dfrac{\sigma}{\sqrt{n}} = \dfrac{1.7}{\sqrt{30}} = 0.3 \right)$.

If we take repeated samples of size 50 from the population, compute the sample mean, $\bar{X}$, for each sample, and plot those sample means, we again get a differently shaped distribution. Sample means, based on samples of size $n = 50$, range from 2.1 to 4.1 and look approximately normally distributed. The mean of all sample means of size $n = 50$ is 3.0, and the standard deviation of the sample means, or the standard error, is 0.2 $\left(\dfrac{\sigma}{\sqrt{n}} = \dfrac{1.7}{\sqrt{50}} = 0.2 \right)$.

When a particular attribute (either an individual measure, x, or sample means of a specific sample size, $\bar{X}$) follows a normal distribution, we can use the standard normal distribution, z, to estimate probabilities. We previously converted individual scores, x, to z scores as: $z = \dfrac{x - \mu}{\sigma}$. If the sample mean, $\bar{X}$, follows a normal distribution (and the central limit theorem supports this for large samples, $n \geq 30$, or for any size samples if we know that the outcome follows a normal distribution in the population), then we convert sample means, $\bar{X}$, to z scores as: $z = \dfrac{\bar{X} - \mu}{\sigma/\sqrt{n}}$. We use this result from the central limit theorem to develop confidence interval estimates.

Confidence Interval Estimates

In practice, we generate confidence interval (CI) estimates, which are ranges of plausible values for a true population parameter based on sample data. Confidence interval estimates for unknown population parameters take the following form: point estimate ± margin of error. The point estimate is our best estimate of the true population parameter (e.g., the sample mean, $\bar{X}$, is the point estimate for the population mean, μ).

The margin of error quantifies sampling variability. The margin of error includes two components—the standard error of the point estimate (which incorporates the variability of the outcome and the sample size) and a z score (if we can assume that the sampling distribution is approximately normal; otherwise, we use an analogous value from another probability distribution—we will say more on this in the next section—Confidence Interval Estimates for a Population Mean or a Population Proportion) that reflects our selected confidence level.

When we estimate CIs, we choose a confidence level, usually 90%, 95%, or 99%. A 95% confidence level is fairly standard in practice, but other levels can be chosen. The confidence level is a probability. It is not the probability that the computed CI contains the true population parameter (a common misunderstanding); the interpretation is a bit more nuanced.

The interpretation can be described as follows. Suppose we choose to construct a 95% CI for an unknown population parameter. If we were to generate all possible 95% CIs for the unknown population parameter, then 95% of these CIs would include, or cover, the true unknown population parameter. In practice, we construct one CI estimate based on data in the sample. It may, or may not, include the true population mean—we can never know for sure.

In the following sections, we illustrate the calculation of CI estimates for means, proportions, and comparisons of means, proportions, and rates. The statistical formulas are presented and illustrated along with appropriate interpretations.

Confidence Interval Estimates for a Population Mean or a Population Proportion

In many situations, we wish to estimate the mean of a continuous outcome in a population (μ) or the proportion with a particular health outcome in a population (p) based on a random sample of participants from that population. **TABLE 2-8** contains the formulas to compute CI estimates for a population mean and for a population proportion.

TABLE 2-8 Confidence Interval Formulas for the Population Mean and Population Proportion

True Population Parameter (Variable Type)	Point Estimate	Sample Size	Confidence Interval Formula
μ (continuous)	$\bar{X}$	$n \geq 30$	$\bar{X} \pm z\dfrac{s}{\sqrt{n}}$
		$n < 30$	$\bar{X} \pm t\dfrac{s}{\sqrt{n}}$, degrees of freedom $(df) = n - 1^c$
p (dichotomous)	$\hat{p}$	$n\hat{p} \geq 5$ and $n(1-\hat{p}) \geq 5^d$	$\hat{p} \pm z\sqrt{\dfrac{\hat{p}(1-\hat{p})}{n}}$

[c] Degrees of freedom (df) refers to the number of independent observations in a sample. When we estimate a mean or a proportion in one sample, df $= n - 1$ once the sample mean or proportion is computed, $n - 1$ of the remaining values are free to vary.

[d] The condition $n\hat{p} \geq 5$ and $n(1-\hat{p}) \geq 5$ translates to having at least 5 participants with the outcome and 5 without the outcome in the study sample. If this condition is not met, then an exact procedure must be used.

Note that in the CI formulas for the population mean, μ, we use the sample standard deviation, s, to estimate the standard error as it is not typical to know the population standard deviation, σ, and not the population mean, μ. If the sample size is large ($n \geq 30$), the central limit theorem supports the use of the standard normal distribution, z. Specifically, z in the CI formula is the z score for the desired confidence level. For a 95% confidence level, $z = 1.96$. Recall in the standard normal distribution (Figure 2-3) that approximately 95% (actually 95.5%) of the z distribution is between -2 and 2 (two standard deviations above and below the mean of 0). To be exact, 95% of the distribution is between -1.96 and 1.96. For a 90% confidence interval, $z = 1.645$, and for a 99% confidence interval estimate, $z = 2.576$. To be more confident that an interval covers the true mean, we need a wider interval. Thus, there is a trade-off between a higher confidence level and the width of the CI. In practice, a 95% confidence level is typically used.

There are many applications where the sample size is small (i.e., $n < 30$). This may be due to highly specific eligibility criteria for the sample, financial or logistical constraints in recruiting and measuring the outcome of interest. If the outcome can be assumed to follow a normal distribution, the population standard deviation is unknown (as it is in most applications) and the sample size is less than 30, then we cannot use a z score in our CI formula and need to use another probability distribution—the student t distribution. The t distribution is similar in form to the standard normal distribution, but takes a slightly different shape depending on the exact (small) sample size. **FIGURE 2-12** shows the standard normal distribution, z, along with t distributions for samples of sizes 5 and 15.

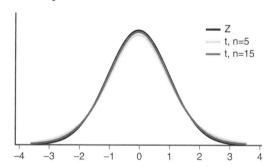

FIGURE 2-12 Standard normal distribution, z, t distributions with sample sizes of $n = 5$ and $n = 15$.

If the sample size is small ($n < 30$), the CI formula to estimate a true population mean involves a value from the t distribution as opposed to z. The t value is a value from the t distribution based on the exact sample size and indexed by degrees of freedom, $\mathrm{df} = n - 1$, for the desired confidence level. The t value can be found from a published t distribution table, from a statistical computing package, or using Excel's t.inv function.[e]

---

e The t.inv function requires two arguments: t.inv(probability, df). The probability is the area under the curve below the desired t value and df is the degree of freedom. Suppose we wish to determine the t value for a 95% confidence interval with df = 14 (i.e., $n = 15$). We specify t.inv(0.975,14) = 2.14.

The CI formula to estimate a population mean involving t can actually be used for any sample size (large or small) because the t value converges to the z value when the sample size is large. Thus, we consistently use the formula with t, as is typical in practice.

The CI for the true proportion is estimated using a z score as is appropriate as long as we have a sample with at least 5 participants with the outcome and at least 5 without. If we have fewer than 5 of each type in the sample, we need alternative procedures, called exact procedures, to generate CI estimates for a population proportion.[14]

EXAMPLE 2-8 illustrates the calculation of CI estimates for continuous and dichotomous variables using the formulas outlined in Table 2-8.

EXAMPLE 2-8

The leading cause of death in the United States is heart disease, and more than 600,000 people die of heart disease every year.[15] The primary risk factors for heart disease are high blood pressure, high cholesterol, and cigarette smoking.[16] We wish to evaluate the extent to which young adults living in Boston might be at risk for heart disease. We recruit a sample of $n = 200$ adults between the ages of 20 and 34 years living in Boston in 2019. For each participant, we measure their systolic and diastolic blood pressure and we collect a blood sample to measure total serum cholesterol. Each participant is also asked whether they currently smoke cigarettes. The sample data are summarized using appropriate descriptive statistics in **TABLE 2-9**. And 95% confidence intervals are computed for each risk factor.

TABLE 2-9 Summary Statistics on Risk Factors for Heart Disease Among $n = 200$ Boston Residents Aged 20–34 Years in 2019

Characteristic	Mean(s) or n (%)	95% CI[f]
Systolic blood pressure, mmHg	121.3 (14.9)	(119.2, 123.4)
Diastolic blood pressure, mmHg	82.4 (8.2)	(81.3, 83.5)
Total serum cholesterol, mg/dL	194.7 (32.6)	(190.2, 199.2)
Current smoking status	25 (12.5)	(7.9, 17.1)

[f] The 95% CI for systolic blood pressure is computed as: $121.3 \pm 1.96 \frac{14.9}{\sqrt{200}}$. The 95% CI for current smoking status is computed as: $0.125 \pm 1.96 \sqrt{\frac{0.125(1-0.125)}{200}}$.

Based on these data, we are 95% confident that the mean systolic blood pressure for all young adults between the ages of 20 and 34 years living in Boston in 2019 is between 119.2 and 123.4 mmHg. We are also 95%

confident that between 7.9% and 17.2% of young adults between the ages of 20 and 34 years living in Boston in 2019 are current smokers.

EXAMPLE 2-9 illustrates the use of the *t* distribution to generate a CI estimate for a population mean when the sample size is small.

EXAMPLE 2-9

We sample $n = 10$ patients between the ages of 20 and 34 years who are free of diabetes and measure their body mass index (BMI). Their mean BMI is 27.3 kg/m² with a standard deviation of 2.10 kg/m². We wish to generate a 95% CI estimate for the true mean BMI among all patients free of diabetes based on this sample.

Because the sample size is small, we must use the CI formula involving *t*. The *t* value for 95% confidence with 9 degrees of freedom is $t = 2.26$.[g] The 95% CI is: $27.3 \pm 2.26 \dfrac{2.1}{\sqrt{10}}$, which is equivalent to 27.3 ± 1.5 kg/m², or (25.8, 28.8). Some investigators report the point estimate ± the margin of error (i.e., 27.3 ± 1.5 kg/m²), whereas others add and subtract the margin of error, and report the upper and lower 95% confidence limits: (25.8, 28.8).

Suppose we change the confidence level to 90%. This changes the *t* value and the margin of error. The *t* value for 90% confidence with 9 df is $t = 1.83$. The 90% CI is $27.3 \pm 1.83 \dfrac{2.1}{\sqrt{10}}$, which is equivalent to 27.3 ± 1.2 kg/m², or (26.1, 28.5). If we change the confidence level to 99%, the *t* value $t = 3.25$ and the 99% CI is $27.3 \pm 3.25 \dfrac{2.1}{\sqrt{10}}$, which is equivalent to 27.3 ± 2.2 kg/m², or (25.1, 29.5). Note that the higher the confidence level, the wider the CI.

Confidence Intervals Comparing Means in Two Independent Groups

Many situations involve a direct comparison of two groups in terms of a continuous outcome. These two groups might be defined by the investigator (e.g., participants randomized to either the intervention or control group in a clinical trial) or based on a particular attribute (e.g., patients with, or without, a family history of disease). When the two groups are physically separate, they are called independent groups, and we compare means by estimating the difference in means between groups. The formula to estimate the difference in population means between two independent groups is: $\left(\bar{X}_1 - \bar{X}_2 \right) \pm t\, S_P \sqrt{\dfrac{1}{n_1} + \dfrac{1}{n_2}}$.

---

g The t value for 95% confidence, df = 9, is computed using Excel: t.inv(0.975,9).

The point estimate for the difference in populations means $(\mu_1 - \mu_2)$ is the difference in sample means $(\bar{X}_1 - \bar{X}_2)$. The margin of error for the difference in means is $t\,S_p \sqrt{\dfrac{1}{n_1} + \dfrac{1}{n_2}}$, which includes a t value to reflect the selected confidence level and is determined based on df $= n_1 + n_2 - 2$. The standard error of the difference in means is $S_p \sqrt{\dfrac{1}{n_1} + \dfrac{1}{n_2}}$, where S_p is the pooled estimate of the common standard deviation $\left(S_p = \sqrt{\dfrac{(n_1 - 1)s_1^2 + (n_2 - 1)s_2^2}{(n_1 + n_2 - 2)}} \right)$.h

EXAMPLE 2-10 illustrates the computation of a CI estimate for the difference in means in two independent groups.

EXAMPLE 2-10

White blood cells (WBCs) are important for the body to fight infection. Cancer patients undergoing chemotherapy treatment often experience a loss of WBCs as the chemotherapy kills both cancerous and infection-fighting cells. A clinical trial is designed to investigate whether there is a difference in WBC count following different routes of administration of the same chemotherapy drugs. Specifically, patients are randomized to receive either a bolus dose (a large dose given over a short period) or continuous intravenous infusion over a prolonged period. WBC count, measured as cells per microliter (μL) of blood, is measured in each patient 48 hours after completion of the assigned treatment. A total of $n = 40$ patients is involved in the trial and they are randomly assigned to competing regimens. The WBC counts are summarized for patients in each treatment regimen in **TABLE 2-10**.

TABLE 2-10 White Blood Cell (WBC) Counts by Treatment Regimen

Regimen	Number of Patients	Mean (Standard Deviation) WBC Count, cells/μL	95% CI for Mean WBC Count, cells/μL
Bolus dose	20	6,234 (957)	(5,600, 6,867)
Continuous intravenous infusion	20	6,708 (1,045)	(6,016, 7,400)

h s_p^2 is computed by taking a weighted average of the two sample variances (s_1^2 and s_2^2) assuming that the variability of the outcome is similar in the comparison groups.

The rightmost column of Table 2-10 contains 95% CI estimates for the mean WBC count in each treatment regimen, considered separately. Our goal is to compare mean WBC counts between the two groups based on the difference in mean WBC counts between regimens. We construct a 95% CI estimate for the difference in means: $(6,234 - 6,708) \pm 2.02(1,002)\sqrt{\dfrac{1}{20} + \dfrac{1}{20}}$.[i]

We can report the CI as −474 ± 640 cells/μL, or we can add and subtract the margin of error to produce a 95% CI for the difference in means (−1,114, 166). Note that our best estimate of the difference in mean WBC counts between treatments is −474 cells/μL with those receiving the bolus treatment having lower mean WBC counts by 474 cells/μL. The 95% CI builds in an estimate of sampling variability and suggests that this difference in mean WBC counts could be anywhere from 1,114 cells/μL lower among those receiving the bolus dose to 166 cells/μL higher among those receiving the continuous intravenous infusion. When the CI for a difference in means includes 0 (which would imply no difference between groups), we conclude that there is no statistically significant difference between the groups as it is possible (with 95% confidence) that the difference is 0, as 0 is included in the CI estimate.

There is another way to compare WBC counts between groups and that is to compare the CIs that we generated for each group separately (see Table 2-10). However, this approach is often misunderstood. When CIs for two independent groups do not overlap, we can be sure that there is a statistical difference between the groups (i.e., the CI for the difference in means will not include 0). However, the opposite is not necessarily true. When CIs for two independent groups overlap, we cannot be sure that there is not a statistical difference between the groups. Thus, if the goal is to compare groups, the recommended approach is to generate a CI estimate for the difference in means in order to draw the appropriate conclusion.

Confidence Intervals Comparing Means in Two Matched or Paired Groups

Some study designs involve taking two measurements for each participant. These measurements might be under two different conditions or at two different points in time. For example, suppose we wish to evaluate the efficacy[j] of a new cream designed to reduce joint pain in patients with arthritis. We recruit a sample of patients with arthritis into the study and have them apply the new cream with active pain-reducing ingredients such as salicylate and capsaicin to one knee and a placebo cream (that looks identical, but does not contain the pain-reducing ingredients) to their other knee. After a specified

---

i   The t value for the 95% CI with df = 20 + 20 − 2 = 38 for the difference in means is computed using Excel with t.inv(0.975, 38).

j   Efficacy refers to the observed effect of treatment or intervention under ideal conditions (e.g., in a controlled clinical trial), compared to effectiveness which is the effect under realistic, real-world conditions.

period of time, each patient reports their pain levels on a continuous scale in each knee. The two measurements are matched by patient.

Another example might involve evaluating the efficacy of a new medication for hypertension. A sample of patients with prehypertension, defined as systolic/diastolic blood pressure between 120/80 and 139/89 mmHg, are recruited to participate in the evaluation. We measure blood pressures in each patient before and after receiving the new medication. The two measurements (pretreatment and posttreatment) are matched by patient.

When we have two matched or paired groups, rather than reporting means for each group separately and computing the difference in means, we instead compute a difference score for each patient. Specifically, we subtract one measurement from the other for each patient to produce a difference or change score—a within-patient difference in outcome under two different conditions or over time. We then summarize these difference or change scores by computing the sample mean of the differences $\left(\bar{X}_d \right)$ and the sample standard deviation of the differences (s_d). Here the subscript d indicates that these summary statistics are based on difference scores, as opposed to raw scores. The CI formula to estimate the mean difference in the population is: $\bar{X}_d \pm t \dfrac{s_d}{\sqrt{n}}$, and the t value is based on $n - 1$ df.

EXAMPLE 2-11 illustrates the computation of a CI estimate to compare means in two matched or paired groups.

EXAMPLE 2-11

Overweight and obesity are leading causes of death in the United States. Suppose we want to estimate the change in BMI in kilograms per meters squared (kg/m²) in a sample of $n = 180$ adults between the ages of 45 and 60 years who are free of cardiovascular disease over the period 2013–2018. Each participant has their BMI measured in 2013 and again in 2018. This is a study with $n = 180$ matched pairs, and we compute difference scores for each participant (subtracting the BMI measured in 2013 from that measured in 2018) and report summary statistics on the differences in BMI over the 5-year period. BMIs at each time point and the changes over time are summarized in **TABLE 2-11**.

TABLE 2-11 Body Mass Index in $n = 180$ Adults 45–60 Years of Age and Free of Cardiovascular Disease

Year	Mean (Standard Deviation) BMI, kg/m²	95% CI for Mean BMI, kg/m²
2013	25.9 (3.9)	(25.3, 26.5)
2018	26.4 (4.4)	(25,8, 27,1)
5 Year Difference	0.5 (2.5)	(0.1, 0.9)

The bottom row of Table 2-11 contains summary statistics and the 95% CI for the mean difference in BMI over time.[k] The point estimate for the within-person change in BMI over time is 0.5 kg/m^2 (i.e., our best estimate of the 5-year change in BMI is 0.5 kg/m^2) and we are 95% confident that the true mean difference in BMI from 2013 to 2018 is between 0.1 and 0.9 kg/m^2. Because this CI for the true mean difference does not include 0 (which would imply no difference over time), we conclude that there is a statistically significant increase in BMI between 0.4 and 0.6 kg/m^2 over the period 2013–2018 as 0 is not included in the CI estimate.

Confidence Intervals Comparing Proportions, Risks, and Rates in Two Independent Groups

Many situations involve a direct comparison of a dichotomous outcome (e.g., diagnosis of autism or the development of a substance use disorder) between two independent groups. The independent groups might be defined by the investigator (e.g., by randomization in a clinical trial) or based on a particular attribute of participants in the study (e.g., age < 30 years vs. 30 years of age and older).

We summarize dichotomous outcomes using proportions, and these proportions might represent the prevalence of a particular health condition (existing cases) or incidence proportions (new cases) in each comparison group. There are other instances where we do not have complete follow-up data to compute the incidence proportion, but we are able to estimate an incidence rate (new cases per a specified time frame) in each comparison group.

To compare proportions, risks, or rates between two independent groups, we can generate absolute or relative comparisons. Absolute comparisons are based on differences, and relative comparisons are based on ratios. Both are acceptable approaches for comparing proportions, risks, and rates in two independent groups, each has advantages and disadvantages, and appropriate interpretation often requires the presentation of both. For example, a relative comparison might suggest that patients treated with *A* have lower risk of infection by 50% (which sounds impressive) compared to patients treated with *B*, but the absolute risk of infection with treatment *A* is 1% compared to 2% with treatment *B*. In order to appropriately interpret the 50% reduction, it is important to know the absolute risks.

EXAMPLE 2-12 describes a study with two independent comparison groups and a dichotomous outcome where the goal is to compare proportions with the outcome of interest between groups.

---

k The CI for the mean difference is computed using $\overline{X}_d \pm t\dfrac{s_d}{\sqrt{n}}$, with df = 179

[t.inv(0.975,179) = 1.97]. Note that we could use $z = 1.96$ in the confidence interval as the sample size is large, but t can be used for any sample size.

EXAMPLE 2-12

A recent study evaluated the association between sleep, anxiety, depression, and obesity among Puerto Rican adolescents aged 10–19 years.[17] **TABLE 2-12** summarizes the numbers of adolescents who reported not getting, and getting, the recommended number of hours of sleep per night (7–9 hours) by obesity status.[l]

TABLE 2-12 Sleep and Obesity Among Puerto Rican Adolescents

	Obese	Not Obese	Total
Not getting recommended hours of sleep	40	102	142
Getting recommended hours of sleep	44	192	236
Total	84	294	378

l Caregiver- or self-reported height and weight were converted into body mass index (BMI) and compared to the CDC's growth charts.[13] In this study, obesity was defined as BMI greater than, or equal to, the 95th percentile for adolescents of the same age and gender.

The proportion of adolescents getting the recommended hours of sleep is estimated at $236/378 = 62.4\%$ (95% CI: 57.4%–67.4%) and the proportion meeting criteria for obesity is estimated at $84/378 = 22.2\%$ (95% CI: 18.2%–26.2%).

We wish to evaluate whether there is a relationship between sleep and obesity. Among those not getting the recommended hours of sleep, $40/142 = 28.2\%$ are obese, compared to $44/236 = 18.6\%$ of those getting the recommended hours of sleep. We now compare the proportions of adolescents who are obese by sleep status using absolute and relative comparisons.

In absolute terms, the point estimate for the difference in proportions of adolescents who are obese in those who do not get the recommended hours of sleep compared to those who do sleep enough is: $28.2\% - 18.6\% = 9.6\%$ (i.e., there is a 9.6 percentage point higher risk of obesity among those who do not get the recommended hours of sleep).[m]

m When computing a difference in proportions or risks, the convention is to subtract the proportion in the control or unexposed group from the proportion in the experimental or exposed group.

In relative terms, the relative risk—also called the risk ratio (RR)—is computed as: 28.6%/18.6% = 1.54.[n] Adolescents who do not get the recommended hours of sleep are 1.54 times as likely to be obese as those who do get enough sleep. There is another way to interpret this ratio. Adolescents who do not get the recommended hours of sleep have a 54% higher risk of being obese compared to those who do sleep enough. Putting this elevation in risk of obesity into context requires knowing that 18.6% of those getting the recommended hours of sleep are obese.

Another relative measure that can be computed with dichotomous outcomes and two independent groups is an odds ratio (OR). When we estimate a proportion or a risk, we divide the number with a particular condition (x) by the total who could experience that condition (i.e., $\hat{p} = x / n$). We demonstrated this for each of the comparison groups. To compute odds, we divide the number with a particular condition (x) by those without the condition ($n - x$). The OR is the ratio of odds in the comparison groups. Here, the OR is computed as: $(40/102)/(44/192) = 0.392/0.229 = 1.71$. Adolescents who do not get the recommended hours of sleep have 1.71 times the odds of being obese compared to those who do get enough sleep. Another way to interpret the OR is: Adolescents who do not get the recommended hours of sleep have 71% higher odds of being obese compared to those who do get enough sleep.

In practice, we aim to estimate the relative risk or RR, as its interpretation is easier than the interpretation of the OR. However, there are some study designs (such as a case control study) where it is not possible to estimate an RR, but calculating an OR is possible.[18] When the risk of outcome is low (i.e., when the outcome is rare), the OR and RR will be similar in magnitude; thus, the OR is a good estimate of the RR when we analyze rare events. We discuss the odds ratio and its use in biostatistical analysis in more detail in Unit 3.

In studies where we measure the number of new cases per person-time, we compute an incidence rate (as described in Unit 1) and we can also compare incidence rates between groups. **TABLE 2-13** displays the CI formulas for differences and ratios of proportions, risks, and rates between two independent groups.

Returning to Example 2-12, we now estimate 95% CIs for the difference in proportions and for the relative risk, or RR. The difference in proportions was estimated as 9.6%. A 95% CI for the difference in proportions is

$$(0.282 - 0.186) \pm 1.96 \sqrt{\frac{0.282(1-0.282)}{142} + \frac{0.186(1-0.186)}{236}}, 0.096 \pm 1.96(0.045),$$

$0.096 \pm 0.088,$ or $(0.008, 0.184)$. Our point estimate for the difference in

---

n When computing a ratio of proportions or risks, the convention is to place the proportion in the experimental or exposed group in the numerator and the proportion in the control or unexposed group in the denominator.

TABLE 2-13 Confidence Interval Formulas for Differences and Ratios of Proportions, Risks, and Rates Between Two Independent Groups

Comparison Type: Effect Measure	True Population Parameter	Point Estimate	Confidence Interval Formula*
Absolute: difference in proportions or risks	$(p_1 - p_2)$	$(\hat{p}_1 - \hat{p}_2)$	$(\hat{p}_1 - \hat{p}_2) \pm Z\sqrt{\dfrac{\hat{p}_1(1-\hat{p}_1)}{n_1} + \dfrac{\hat{p}_2(1-\hat{p}_2)}{n_2}}$
Relative: risk ratio	$RR = \dfrac{p_1}{p_2} = \dfrac{x_1/n_1}{x_2/n_2}$	$\widehat{RR} = \dfrac{\hat{p}_1}{\hat{p}_2}$	Compute 95% CI for $\ln(RR)$ $\ln(\widehat{RR}) \pm Z\sqrt{\dfrac{1-\hat{p}_1}{n_1\hat{p}_1} + \dfrac{1-\hat{p}_2}{n_2\hat{p}_2}}$ Exponentiate limits to get CI for RR
Relative: odds ratio	$OR = \dfrac{o_1}{o_2} = \dfrac{x_1/(n_1-x_1)}{x_2/(n_2-x_2)}$	$\widehat{OR} = \dfrac{\hat{o}_1}{\hat{o}_2}$	Compute 95% CI for $\ln(OR)$ $\ln(\widehat{OR}) \pm Z\sqrt{\dfrac{1}{x_1} + \dfrac{1}{(n_1-x_1)} + \dfrac{1}{x_2} + \dfrac{1}{(n_2-x_2)}}$ Exponentiate limits to get CI for OR

(continues)

TABLE 2-13 Confidence Interval Formulas for Differences and Ratios of Proportions, Risks, and Rates Between Two Independent Groups *(continued)*

Comparison Type:

Effect Measure	True Population Parameter	Point Estimate	Confidence Interval Formula*
Relative: rate ratio	$IRR = \dfrac{IR_1}{IR_2} = \dfrac{x_1/T_1}{x_2/T_2}$	$\widehat{IRR} = \dfrac{\widehat{IR_1}}{\widehat{IR_2}}$	Compute 95% CI for $\ln(IRR)$ $$\ln\left(\widehat{IRR}\right) \pm Z\sqrt{\dfrac{1}{x_1} + \dfrac{1}{x_2}}$$ Exponentiate limits to get CI for IRR

*These CI formulas involve z scores and assume that there are at least 5 participants with the outcome and at least 5 without the outcome in each comparison group.

*The CI formulas for differences in proportions or risks, the RR, OR, and rate ratio use the following notation:

	Number with Outcome	Number Free of Outcome	Total Number of Participants	Total Person-Time
Group 1 (experimental, exposed)	x_1	$n_1 - x_1$	n_1	T_1
Group 2 (control, unexposed)	x_2	$n_2 - x_2$	n_2	T_2

proportions of obese children comparing those who do not get the recommended hours of sleep and who do get the recommended hours of sleep is 9.6% with a 95% CI (0.8%, 18.4%). Because this CI for the difference in proportions does not include 0 (which would imply no difference), we can conclude that a statistically significantly higher proportion of children who do not get the recommended hours of sleep are obese compared to those who do get the recommended hours of sleep.

The RR was estimated as 1.51. A 95% CI for the RR is computed as follows. First we estimate a 95% CI for the ln(RR): $\ln(1.51) \pm 1.96\sqrt{\dfrac{1-0.282}{142(0.282)} + \dfrac{1-0.186}{236(0.186)}} = 0.412 \pm 1.96(0.191), 0.412$ ± 0.374, or $(0.038, 0.786)$. The final step is to exponentiate each of these limits to produce the 95% CI for the RR: $[\exp(0.038), \exp(0.786)]$, or $(1.04, 2.19)$. Our point estimate for the RR is 1.51 with a 95% CI $(1.04, 2.19)$. Because this CI for the RR does not include 1 (which would imply no difference), we conclude that a statistically significantly higher proportion of children who do not get the recommended hours of sleep are obese compared to those who do get the recommended hours of sleep.

Confidence Intervals Comparing Proportions and Risks in Two Matched or Paired Groups

In this section, we discuss applications where we have a dichotomous outcome (e.g., presence or absence of disease) and matched or paired groups. Specifically, we classify participants twice in terms of a dichotomous outcome. These classifications might be under two different conditions or at two different points in time. Again, we can generate absolute and relative comparisons of matched proportions using CIs for the difference in matched proportions and the ratio of matched proportions, respectively.

EXAMPLE 2-13 illustrates the computation of a CI estimate to compare proportions in two matched or paired groups.

EXAMPLE 2-13

Two different throat cultures (culture *A* and culture *B*) that test for strep throat are compared in a sample of *n* = 100 patients. Two swabs are collected from each patient, and one is tested in culture *A* and the other in culture *B*. After a few days in the laboratory, each culture is classified as positive for bacterial infection or not. The data are summarized in **TABLE 2-14**.

(continues)

EXAMPLE 2-13   (*continued*)

TABLE 2-14 Summarizing Dichotomous Outcomes in Two Matched or Paired Groups

	Culture A–Positive	Culture A–Negative	Total
Culture B–Positive	25	12	37
Culture B–Negative	21	42	63
Total	46	54	100

Based on culture A, there are $25 + 21 = 46$ positive results and based on culture B, there are $25 + 12 = 37$ positive results. The goal is to compare the two cultures to see whether they produce different results. In this study, 25 participants test positive on both cultures and 42 participants test negative on both cultures. These $25 + 42 = 67$ participants who are classified in the same way on both culture A and culture B are called concordant pairs. The data from these participants actually do not help us understand whether there are any differences between the cultures. We are most interested in the discordant pairs (i.e., the data from the 12 and 21 participants who were classified differently).

TABLE 2-15 displays the CI formulas for the difference in matched proportions and for ratios of matched proportions.[19,20]

Returning to Example 2-13, the 95% CI for the difference in matched proportions is computed as: $\dfrac{(c-b)}{n} \pm z\dfrac{\sqrt{b+c}}{n}$, $\dfrac{(21-12)}{100} \pm 1.96\dfrac{\sqrt{12+21}}{100}$, $0.09 \pm 1.96(0.057)$, 0.09 ± 0.112, or $(-0.022, 0.202)$. Our point estimate for *the difference in proportions of positive cultures is 0.09. Thus, the propor-*tion of positive tests with culture A is 9 percentage points higher than the proportion of positive tests with culture B, with a 95% CI $(-2.2\%, 20.2\%)$. Because this CI for the difference in proportions includes 0 (which implies no difference), we cannot conclude that there is a statistically significant difference in the proportion of positive tests with culture A compared to the proportion of positive tests with culture B.

The RR is estimated as $\widehat{RR} = \dfrac{25+21}{25+12} = 1.24$. A 95% CI for the RR is computed as follows. First, we estimate a 95% CI for the ln(RR): $\ln(1.24) \pm 1.96$
$\sqrt{\dfrac{12+21}{(25+12)\times(25+21)}} = 0.215 \pm 1.96(0.139), 0.215 \pm 0.272,$ or $(-0.057, 0.487)$.
The final step is to exponentiate each of these limits to produce the 95% CI

TABLE 2-15 Confidence Interval Formulas for Differences and Ratios of Matched Proportions

Comparison Type: Effect Measure	True Population Parameter	Point Estimate	Confidence Interval Formula*
Absolute: difference in matched proportions or risks	$\left(p_1 - p_2\right)$	$\left(\hat{p}_1 - \hat{p}_2\right) = \dfrac{c - b}{n}$	$\dfrac{(c-b)}{n} \pm z\dfrac{\sqrt{b+c}}{n}$
Relative: risk ratio for matched proportions	$RR = \dfrac{p_1}{p_2}$	$\widehat{RR} = \dfrac{\hat{p}_1}{\hat{p}_2} = \dfrac{a+c}{a+b}$	Compute 95% CI for $\ln(\widehat{RR})$ $$\ln\left(\widehat{RR}\right) \pm Z\sqrt{\dfrac{b+c}{(a+b)\times(a+c)}}$$ Exponentiate limits to get CI for RR
Relative: odds ratio for matched proportions	$OR = \dfrac{o_1}{o_2}$	$\widehat{OR} = \dfrac{\hat{o}_1}{\hat{o}_2} = \dfrac{c}{b}$	Compute 95% CI for $\ln(\widehat{OR})$ $$\ln\left(\widehat{OR}\right) \pm Z\sqrt{\dfrac{1}{b} + \dfrac{1}{c}}$$ Exponentiate limits to get CI for OR

(continues)

TABLE 2-15 Confidence Interval Formulas for Differences and Ratios of Matched Proportions

*These CI formulas involve z scores and assume that there are at least 5 participants with the outcome and at least 5 without in each comparison group.

*The CI formulas for differences in matched proportions and ratios of matched proportions use the following notation:

Condition 1

		+	−
Condition 2	+	a	b
	−	c	d

$$\hat{p}_1 = \frac{a+c}{n}$$

$$\hat{p}_2 = \frac{a+b}{n}$$

$$n$$

(*continued*)

for the RR: [exp(−0.057), exp(0.487)], or (0.95, 1.63). Our point estimate for the RR is 1.24 with a 95% CI (0.95, 1.63). Because this CI for the RR includes 1 (which implies no difference), we cannot conclude that there is a statistically significant difference in the proportion of positive tests with culture A compared to the proportion of positive tests with culture B.

Just as with independent proportions, the preferred ratio estimate is the RR. However, it is possible to estimate an OR as: $\widehat{OR} = \dfrac{21}{12} = 1.75$. A 95% CI for the OR is computed as follows. First, we estimate a 95% CI for the $\ln(OR)$: $\ln(1.75) \pm 1.96\sqrt{\dfrac{1}{12} + \dfrac{1}{21}} = 0.560 \pm 1.96(0.362), 0.560 \pm 0.710$, or $(-0.150, 1.270)$. The next, and final, step is to exponentiate each of these limits to produce the 95% CI for the OR: [exp(−0.150), exp(1.270)], or (0.86, 3.56). Our point estimate for the OR is 1.75 with a 95% CI (0.86, 3.56). Because this CI for the OR includes 1 (which implies no difference) we cannot conclude that there is a statistically significant difference in the proportion of positive tests with culture A compared to the proportion of positive tests with culture B.

▶ 2.5 Tests of Hypothesis for Means and Proportions

Statistical inference includes two broad areas, estimation (outlined in Section 2.4) and tests of hypothesis. In applications involving estimation, we generate point and CI estimates for unknown population parameters such as the population mean (μ), population proportion (p), the difference in means or proportions, $\mu_1 - \mu_2$ or $p_1 - p_2$, respectively.

In hypothesis testing applications, we first formulate a hypothesis or statement of what we believe to be true in the population. For example, we might hypothesize that patients who engage in 15 minutes of high-intensity exercise per day have lower systolic blood pressures on average than those who do not exercise, or that adolescents who vape are at higher risk for heart attack than adolescents who do not vape. These are examples of research hypotheses. For each research hypothesis, we additionally specify a null hypothesis, which essentially reflects the opposite situation of no association, no difference, or no effect. The null hypothesis is the comparison hypothesis. In the example evaluating the impact of exercise on blood pressure, the null hypothesis would be that there is no difference in blood pressure between those who engage in 15 minutes of high-intensity exercise per day compared to those who do not exercise. In the vaping example, the null hypothesis would be that there is no difference in heart attack risk between those who vape and those who do not vape.

Once the hypotheses are specified, we then use sample data to "test" the hypotheses. The test involves making a judgment as to whether the null hypothesis (i.e., the no association or no difference scenario) is likely true given the sample data or not, appropriately accounting for uncertainty. There are many different statistical procedures for testing hypotheses, just as there were different formulas to produce CI estimates, but all follow the same general approach, which we organize into four steps. Before applying the four steps to conduct tests of hypothesis, we first specify the outcome of interest (e.g., systolic blood pressure, incident disease) and identify its type (e.g., continuous, dichotomous) as this information affects how the steps are applied in the test of hypothesis. With continuous outcomes, tests of hypothesis are generally focused on means. With dichotomous outcomes, tests of hypothesis are generally focused on proportions. We must also specify the number of groups being compared and whether these comparison groups are independent, matched, or paired. These considerations (outcome type, number of comparison groups, relationships among the comparison groups) dictate the specifics of the test of hypothesis. We now outline the steps involved in tests of hypotheses in general terms and then illustrate their use in specific applications.

1. Specify the null (H_0) and research (H_1) hypotheses and the level of significance (α).

   Generally, the research hypothesis is specified first as it reflects what the investigator believes to be true about the population parameter(s). For example, the mean of a continuous outcome among treated patients is higher than the mean among untreated patients, the risk of disease is lower among treated patients compared to those who are untreated, the mean of a continuous outcome among patients receiving a behavioral intervention is different than the mean among those not receiving the intervention. The investigator chooses the appropriate research hypothesis based on their knowledge of the clinical or content area. The investigator might hypothesize a specific direction of effect (an increase or a decrease) or a difference in effect (i.e., an association). These are called one-sided and two-sided tests, respectively, and are illustrated next through examples. The null hypothesis is the competing hypothesis and reflects the no-difference, no-change, or no-association situation.

   The second component of step 1 involves specifying the level of significance to be used in the test of hypothesis. The level of significance, denoted as α, is the probability of rejecting the null hypothesis in favor of the research hypothesis when the null hypothesis is actually true. The level of significance is a probability (the probability of a specific error) called a Type I error. Usually, a 5% level of significance is used, just as 95% was a typical confidence level, although other levels can be selected (e.g., 1%, 10%).

2.   Compute the test statistic based on sample data.

Once a sample is selected from the population, descriptive statistics are computed on the outcome of interest (e.g., for a continuous outcome, we compute the sample size, n, the sample mean, $\overline{X}$, and the sample standard deviation, s, and for a dichotomous outcome, we compute the sample size, n, and the sample proportion, $\hat{p}$) and then summarized into a single value, called the test statistic. The form of the test statistic is specific to the variable type of the outcome of interest (e.g., continuous, dichotomous) and depends on the number of groups being compared and whether those groups are independent, matched, or paired. In general, the test statistic summarizes what is observed in the sample relative to what would be expected if the null hypothesis were true, accounting for sampling variability.

3.   Determine statistical significance of the data.

The statistical significance of the data is determined by computing a p-value, which is the probability of observing a test statistic (which summarizes the sample data) as, or more extreme than, that observed. The p-value allows the investigator to judge whether the data support the research hypothesis or not.

***p*-values.** The statistical significance of a test of hypothesis is summarized with a p-value which quantifies how "incompatible the data are with a specified statistical model."[21] A smaller p-value, specifically a p-value smaller than the specified level of significance (e.g., 0.05), provides statistical evidence that the sample data are incompatible with the null hypothesis. **EXAMPLE 2-14** illustrates how a p-value is computed. In practice, we use a statistical computing package or a program like Excel to compute p-values for specific tests of hypothesis.

EXAMPLE 2-14

Suppose we set up a test of hypothesis to test the research hypothesis that the population mean in a specific population is higher than a reported mean in a comparison population. We summarize the sample data into a z statistic and find $z = 2.5$ (details on test statistics for particular tests are outlined below). We then determine how likely it is to observe such a value (i.e., $z = 2.5$) under the assumed statistical model (e.g., that data follow a normal distribution with a particular mean). If the null hypothesis is true (i.e., the mean in the test population is equal to the mean in the comparison population), then any value of z in **FIGURE 2-13** is theoretically possible. We observe $z = 2.5$. How likely is this value if the null hypothesis is true?

(continues)

EXAMPLE 2-14   (*continued*)

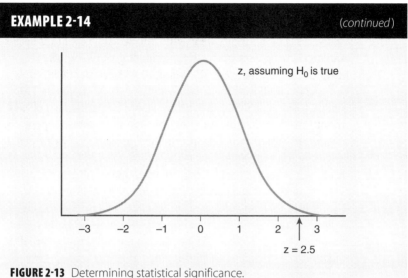

FIGURE 2-13 Determining statistical significance.

The *p*-value is the probability of observing a *z* statistic as, or more extreme than, that observed. Specifically, $P(z > 2.5) = 0.0006$.[o] The threshold for determining statistical significance is the level of significance, α. Assuming that we specify $\alpha = 0.05$, we reject the null hypothesis in favor of the research hypothesis because the *p*-value ($p = 0.0006$) is less than 0.05.

If we had hypothesized that the mean in the population of interest was different from the mean in the comparison population, then the *p*-value would be computed as $2 \times P(z > 2.5) = 2 \times 0.006 = 0.012$. Once again, we would reject the null hypothesis in favor of the research hypothesis as the *p*-value ($p = 0.012$) is less than 0.05. In practice, one research hypothesis is specified and it is often a two-sided research hypothesis suggesting a difference or association, as opposed to a specific direction.

4.   Summarize statistical and clinical or practical significance of the results.

The fourth, and final, step is a summary of the test of hypothesis and involves assessing the statistical and clinical (or practical) significance of the results, and both are important. Statistical significance is determined in Step 3. Clinical or practical significance is determined by the investigator, is judged by their interpretation as to whether the data meaningfully support the research hypothesis or not, and may (or may not) agree with the assessment of statistical significance.

o   The norm.s.dist function in Excel requires two arguments: norm.s.dist(*z*, cumulative). Specifying cumulative = 1 returns the probability below *z*. To compute $P(z > 2.5)$, we specify $1 - \text{norm.s.dist}(2.5, 1)$.

Tests of hypothesis might be highly statistically significant (e.g., $p < 0.0001$) but have little clinical or practical impact, or vice versa. For example, we might find that the risk of disease is statistically different in two independent groups, but the absolute risks of disease are 0.02% versus 0.03%, respectively. How can such a small difference reach statistical significance? This typically happens when the sample size is large. In contrast, we might find that the absolute risks of disease in two independent groups are 12% versus 3%, which seem meaningfully different. However, a test of hypothesis might find $p = 0.354$, which would not be statistically significant as the p-value is not less than 5% (assuming we use the standard criterion of 5%). How can a 4-fold difference in risks not be statistically significant? Most typically, this type of discrepancy occurs when the sample size is small. Thus, it is always important to consider both statistical and clinical (or practical) significance in summarizing tests of hypothesis.

Errors in Statistical Tests

There are two types of errors that can be made with statistical tests of hypothesis. A Type I error occurs when we reject the null hypothesis in favor of the research hypothesis when the null hypothesis is true. The probability of a Type I error is equal to the level of significance, α. When we run a test of hypothesis, we choose levels of significance of 1%, 5%, or 10%, with 5% being most commonly used. Thus, we are reasonably protected against a Type I error.

A Type II error occurs when we fail to reject the null hypothesis when the research hypothesis is true. The probability of a Type II error is more complicated to quantify and is related to the sample size, the true effect if the research hypothesis is true, and the level of significance, α. When we run a test of hypothesis and fail to reject the null hypothesis in favor of the research hypothesis, there may be a high probability of committing a Type II error (defined as β). Unfortunately, we cannot simply specify a probability of a Type II error, β, to not exceed 5% or 10% as this probability depends on a number of other factors. Often, we frame our thinking around a Type II error in terms of statistical power, which is defined as the probability of rejecting a null hypothesis when it is actually false, $1 - \beta$. A better test is one with higher statistical power and lower probability of Type II error, which occurs with a larger sample size. We will discuss the issue of statistical power in more detail through examples.

In the following sections, we work through the details of specific tests of hypothesis, following this four-step approach, summarizing the statistical and clinical (or practical) significance of results.

Hypothesis Tests for a Population Mean or a Population Proportion

In some applications, we wish to test whether a population mean of a continuous outcome, or a population proportion of a dichotomous outcome, is statistically significantly higher, lower, or different from a known value. The known value might be based on historical data or derived from a different population.

These tests are called one-sample tests and the null hypothesis reflects the situation where the mean or proportion in the population of interest is equal to the known mean or proportion. The research hypothesis specifies the hypothesized effect. Specifically, that the mean or proportion in the population of interest is higher, lower, or different from the known mean or proportion, depending on what the investigator believes to be true, and this research hypothesis is set up prior to the collection of any data. Most tests consider the two-sided research hypothesis, specifically that the mean or proportion is different from the known mean or proportion, and set a 5% level of significance.

The test statistics for a one-sample test for a population mean or a population proportion are outlined in **TABLE 2-16** and p-values are calculated by computing probabilities of observing test statistics as, or more extreme than, those observed using z and t probability distributions.

TABLE 2-16 Test Statistics for One-Sample Tests for Population Mean or Population Proportion

True Population Parameter (Variable Type)	Hypotheses	Test Statistic
μ (continuous)	$H_0: \mu = \mu_0{}^*$ $H_1: \mu > \mu_0,$ $\mu < \mu_0,$ or $\mu \neq \mu_0$	$t = \dfrac{\bar{X} - \mu}{s/\sqrt{n}},$ degrees of freedom (df) $= n - 1$
p (dichotomous)	$H_0: p = p_0{}^*$ $H_1: p > p_0,$ $p < p_0,$ or $p \neq p_0$	$Z = \dfrac{\hat{p} - \hat{p}_0}{\sqrt{\dfrac{\hat{p}(1 - \hat{p})}{n}}}$ [p]

$^*\mu_0$ and p_0 are the known mean and proportion, respectively.
[p] Appropriate use requires $n\hat{p} \geq 5$ and $n(1 - \hat{p}) \geq 5$, which translates to having at least 5 participants with the outcome and 5 without in the study sample. If this condition is not met, then an exact procedure must be used.

EXAMPLE 2-15 illustrates the test of hypothesis procedure for a population mean.

EXAMPLE 2-15

Excessive intake of added sugars (i.e., sugars or syrups added to food during processing or preparation) is linked to other unhealthy behaviors and varies by age, sex, race/ethnicity, geography, and socioeconomic status.[22] Current dietary guidelines recommend that added sugars be kept to less than 10% of total caloric intake.[23] We wish to test whether Boston adolescents are adhering to these recommendations or not. Specifically, we wish to test whether the percentage of calories due to added sugars consumed by Boston adolescents is different from the recommended 10%.

A sample of $n = 400$ adolescents aged 12–15 years are enrolled in a study that gathers data on daily food intake through detailed food diaries. The specific foods consumed are analyzed and nutrients are quantified. Total calories consumed per day and calories from added sugars are computed for each participant, as is the percentage of calories from added sugars. In the sample, Boston adolescents consumed a mean of 2,150 calories per day with a mean of 15.4% and a standard deviation of 32.6% of calories from added sugars. Is there statistical evidence of a difference from the recommended 10% of calories from added sugars among Boston adolescents?

1. Specify hypotheses and choose the level of significance.

    H_0: The mean percentage of calories from added sugars is 10% among Boston adolescents ($\mu = 10\%$).

    H_1: The mean percentage of calories from added sugars is different than 10% among Boston adolescents ($\mu \neq 10\%$).

    $\alpha = 0.05$

2. Compute test statistic based on sample data.

    The t statistic is computed as: $t = \dfrac{\bar{X} - \mu}{s/\sqrt{n}} = \dfrac{15.4 - 10}{32.6/\sqrt{400}} = 3.31.$

3. Determine statistical significance of the data.

    This is a two-sided test (the research hypothesis specifies a difference as opposed to a direction), so the p-value is computed as $2 \times P(t > 3.31) = 0.001$.[q] Because we chose $\alpha = 0.05$, we reject the null hypothesis in favor of the research hypothesis as the p-value ($p = 0.001$) is less than 0.05

4. Summarize statistical and clinical or practical significance of the results.

    Among Boston adolescents, 15.4% of total calories consumed are from added sugars. This is statistically significantly different, $p = 0.001$, from the recommended 10%. This is an example where the observed mean (15.4%) is meaningfully and statistically significantly different from the reported mean of 10%.

---

q The t.dist.2t function in Excel computes two-sided p-values and requires two arguments: t.dist.2t($|t|$, df). To compute the two-sided p-value with a test statistic of $t = 3.31$, and df $= 400 - 1 = 399$, we specify t.dist.2t(3.31,399).

Hypothesis Tests Comparing Means in Two Independent Groups

Perhaps the most widely applied statistical test is the test to compare means in two independent groups. The groups might be defined by the investigator (e.g., participants randomized to receive a behavioral intervention or not) or based on a particular condition or attribute (e.g., patients with a family history of disease or not, patients who served time in the military or not). The outcome of interest is the difference in population means between the two independent groups.

In two independent samples tests, the null hypothesis reflects the situation where there is no difference in population means or where the two population means are equal. The research hypothesis specifies the hypothesized effect. Specifically, that one mean is higher than the other or that they are different, depending on what the investigator believes to be true, and this research hypothesis is set up prior to the collection of any data. Most tests consider the two-sided research hypothesis, specifically that the two population means are different, and set a 5% level of significance.

The test statistic for the two independent samples test for means is:

$t = \dfrac{\bar{X}_1 - \bar{X}_2}{S_P\sqrt{\dfrac{1}{n_1} + \dfrac{1}{n_2}}}$, which has df $= n_1 + n_2 - 2$ and where s_p is the common

standard deviation based on the pooled, weighted average of the standard

deviations in the comparison groups: $S_P = \sqrt{\dfrac{(n_1 - 1)s_1^2 + (n_2 - 1)s_2^2}{n_1 + n_2 - 2}}$. p-values

are calculated by computing probabilities of observing a test statistic as, or more extreme than, that observed using the t probability distribution.

EXAMPLE 2-16 illustrates the test of hypothesis procedure to compare means in two independent groups.

EXAMPLE 2-16

Maintaining weight loss is a difficult task for most people, with approximately 20% of people maintaining a 10% weight loss after 12 months.[24] A clinical trial investigates whether in-person compared to online behavioral interventions for weight-loss maintenance produce different results. A total of 150 participants who successfully lost 10% or more of their body weight agree to enroll and are randomized to participate in either the in-person or online program formats. Both programs include support and monitoring of diet, physical activity, sleep, and stress management. Each participant has their weight and height measured at study start, and their weight is monitored on a monthly basis for 1 year. Weights are converted to BMI scores (weight in kilograms/height in meters squared). Change in BMI (BMI at 12 months – BMI

(continues)

EXAMPLE 2-16 (*continued*)

at study start) is measured for each participant and summarized in **TABLE 2-17** according to the intervention program format.

TABLE 2-17 Change in BMI over 12 Months by Intervention Format

Intervention Format	Number of Participants	$\bar{X}(s)$ Change in BMI over 12 Months (kg/m²)
In-person	75	5.0 (8.7)
Online	75	2.8 (6.9)

We wish to test whether there is a difference in change in BMI over 12 months in participants assigned to in-person compared to online weight-loss maintenance programs.

1. Specify hypotheses and choose the level of significance.

    H_0: There is no difference in mean change in BMI over 12 months in participants assigned to in-person compared to online weight-loss maintenance programs ($\mu_1 = \mu_2$).

    H_1: There is a difference in mean change in BMI over 12 months in participants assigned to in-person compared to online weight-loss maintenance programs ($\mu_1 \neq \mu_2$).

    $\alpha = 0.05$

2. Compute test statistic based on sample data.

    The first step in computing the t statistic involves computing the common standard deviation, s_p, based on the pooled, weighted average of the standard deviations in the comparison groups:

    $$S_p = \sqrt{\frac{(n_1 - 1)s_1^2 + (n_2 - 1)s_2^2}{n_1 + n_2 - 2}} = \sqrt{\frac{(74)\,8.7^2 + (74)\,6.9^2}{75 + 75 - 2}} = 7.9.$$

    And next, the test statistic:

    $$t = \frac{\bar{X}_1 - \bar{X}_2}{S_p\sqrt{\dfrac{1}{n_1} + \dfrac{1}{n_2}}} = \frac{5.0 - 2.8}{7.9\sqrt{\dfrac{1}{75} + \dfrac{1}{75}}} = 1.71.$$

3. Determine statistical significance of the data.

    This is a two-sided test (the research hypothesis indicates a difference), so the p-value is computed as $2 \times P(t > 1.71) = 0.089$.[r] At $\alpha = 0.05$, we do not reject the null hypothesis as the p-value ($p = 0.089$) is not less than 0.05.

---

r   The t.dist.2t function in Excel computes two-sided p-values and requires two arguments: t.dist.2t($|t|$, df). To compute the two-sided p-value with a test statistic of $t = 1.71$ and df $= 75 + 75 - 2 = 148$, we specify t.dist.2t(1.71,148).

4. Summarize statistical and clinical or practical significance of the results.

    Participants assigned to the in-person program format gained a mean of 5.0 kg/m² over 12 months compared to 2.8 kg/m² among those assigned to the online program format. This translates to a difference of 2.2 kg/m², about 5.0 lb difference for participants of the same height. This difference may be a meaningful difference, but fails to reach statistical significance.

    When we fail to reject a null hypothesis, it may be that the study is underpowered, often due to a small sample size. Thus, we summarize the test by stating that we did not have sufficient evidence to show that the mean changes in BMI over 12 months were different between program formats as opposed to concluding that the mean changes in BMI are equal between program formats.

Hypothesis Tests Comparing Means in Two Matched or Paired Groups

Some applications involve matched or paired groups where two measurements are taken on each participant either serially in time (e.g., before and after an intervention) or under two different experimental conditions. When we have two matched or paired groups, we compute difference scores for each participant and analyze the mean difference in the population. Note that in Example 2-16 we computed changes in BMI over time, but the key comparison of interest was the difference in changes in BMI between intervention program formats—two independent groups. It is extremely important to recognize the difference between two independent groups versus two matched or paired groups as the statistical procedures to analyze the data are quite different.

In two matched or paired sample tests, the null hypothesis reflects the situation where there is no difference or change over time, no difference or change under different experimental conditions, or a population mean difference of 0. The research hypothesis specifies the hypothesized effect. Specifically, it specifies that the mean difference is positive, negative, or different from 0, depending on what the investigator believes to be true and this research hypothesis is set up prior to the collection of any data. Most tests consider the two-sided research hypothesis, specifically that the population mean difference is not 0, and set a 5% level of significance.

The test statistic for the two matched or paired samples test for the population mean difference is: $t = \dfrac{\bar{X}_d - \mu_d}{s_d / \sqrt{n}}$, which has df $= n - 1$, and p-values are calculated by computing probabilities of observing a test statistic as, or more extreme than, that observed using the t probability distribution.

EXAMPLE 2-17 illustrates the test of hypothesis procedure to compare means in two matched or paired groups.

EXAMPLE 2-17

E-cigarette use among adolescents continues to increase yet the adverse health effects of e-cigarettes are not well understood, particularly among this demographic. E-cigarettes contain nicotine and are therefore addictive, they irritate the lungs, and have been shown to elevate heart rate and blood pressure.[25] An investigator wants to evaluate the impact of a new educational program to inform adolescents of the health risks associated with e-cigarette use and they aim to show an increase in knowledge of specific health risks associated with e-cigarette use following completion of the program.

A total of $n = 20$ adolescents agree to participate in the evaluation. Each takes a self-administered test to evaluate their knowledge of the health risks associated with e-cigarettes before and after completing the educational program. Tests are scored from 0 to 100 with higher scores indicative of better knowledge of health risks, and are summarized in **TABLE 2-18**.

TABLE 2-18 Knowledge of Risks Associated with e-Cigarette Use Before and After Educational Intervention

$\overline{X}(s)$ Knowledge Scores Before Intervention	$\overline{X}(s)$ Knowledge Scores After Intervention	$\overline{X}_d(s_d)$ Difference in Knowledge Scores
52.5 (8.6)	61.3 (9.2)	8.8 (4.9)

Difference scores for each participant are computed by subtracting the initial test score from the score after completing the evaluation and represents the increase in test scores. Here, we wish to test whether there is a statistically significant increase in test scores after completing the educational intervention. (Had the investigator subtracted the final score from the initial score, the research hypothesis would specify the opposite effect.)

1. Specify hypotheses and choose the level of significance.

   H_0: There is no difference in mean test scores before versus after the educational intervention ($\mu_d = 0$)

   H_1: The mean test score following the educational intervention is higher than the mean test score before the intervention (equivalent to specifying that the mean difference is greater than 0, given the way in which difference scores are computed, $\mu_d > 0$).

   $\alpha = 0.05$

2. Compute test statistic based on sample data.

   The t statistic is computed as: $t = \dfrac{\overline{X}_d - \mu_d}{s_d / \sqrt{n}} = \dfrac{8.8 - 0}{4.9 / \sqrt{20}} = 8.03.$

3. Determine statistical significance of the data.

    This is a one-sided test (the research hypothesis indicates an improvement in test scores following the educational intervention), so the *p*-value is computed as $P(t > 8.03) = 0.000000079$.[s] Usually, *p*-values less than 0.0001 are reported as $p < 0.0001$. Thus, at $\alpha = 0.05$, we reject the null hypothesis as the *p*-value ($p < 0.0001$) is less than 0.05.

4. Summarize statistical and clinical or practical significance of the results.

    Before the educational intervention, the mean test score on the knowledge test was 52.5, suggesting substantial room for improvement. After completing the educational intervention, the mean test score rose to 61.3. The mean improvement was 8.8 points, which was highly statistically significant at $p < 0.0001$. It is not clear whether this level of improvement is meaningful (i.e., would investigators have hoped for a more profound impact?) Only the investigators, with more content expertise, could comment on the clinical or practical significance of the result.

Hypothesis Tests Comparing Means in More than Two Groups

It can be of interest to compare more than two independent groups in terms of a continuous outcome. Again, groups might be defined by the investigator (e.g., participants randomized to one of several [more than two] competing treatments in a clinical trial), or based on a particular condition or attribute (e.g., patients aged under 30 years, between 30 and 45 years, or over 45 years). If the outcome of interest is a continuous measure, and the goal is to compare the means among the groups, then the test of hypothesis procedure to compare more than two independent means is analysis of variance (ANOVA).

In ANOVA, the null hypothesis reflects the situation where the population means are all equal. The research hypothesis reflects the situation where the population means are not all equal. There is no direction of effect specified in the research hypothesis for ANOVA. If any two means are not equal, then the research hypothesis is considered to be true.

The test statistic for ANOVA gets a bit more complicated than those described so far as there are several (more than two) sample means, sample standard deviations, and sample sizes to summarize into a single test statistic. Investigators typically use a statistical computing package to compute a test statistic for ANOVA which takes the form of an *F* statistic, assuming

---

s   The t.dist.rt function in Excel computes the *p*-value for an upper tailed test and requires two arguments: t.dist.rt(*t*, df). To compute the one-sided *p*-value with a test statistic of $t = 8.03$ and $df = 20 - 1 = 19$, we specify t.dist.rt(8.03,19).

that the null hypothesis is true. The F statistic for ANOVA is computed by taking the ratio of a measure of variability between groups to a measure of variability within groups. Specifically, the test statistic for ANOVA is:

$$F = \frac{\sum_{j=1}^{k} n_j \left(\bar{X}_j - \bar{X} \right)^2 \Big/ (k-1)}{\sum_{j=1}^{k} \left(n_j - 1 \right) s_j^2 \Big/ (N-k)}, \text{ where } n_j \text{ represents the sample size in the } j\text{th}$$

group ($j = 1, 2, \ldots, k$) and k represents the number of independent groups ($k > 2$), $\bar{X}_j$ is the sample mean in the jth group, $\bar{X}$ is the overall mean, and N represents the total number of observations in the analysis. Here, N does not refer to a population size, but rather the sum of the sample sizes across all independent groups (i.e., $N = n_1 + n_2 + n_3 + \ldots + n_k$). The F statistic is an extension of the t-statistic we used for testing the equality of two independent group means. The F statistic is always positive and larger values suggest greater differences between groups, accounting for variability within groups, and leads to statistical significance.

Statistical significance is summarized in a p-value, specifically the p-value is determined by $P(F > F_{\text{test statistic}})$, where $F_{\text{test statistic}}$ is the observed test statistic. The F distribution is indexed by two degrees of freedom; df_1 is defined as the number of groups -1 ($k - 1$) and df_2 is defined as the total sample size minus the number of groups ($N - k$).

EXAMPLE 2-18 illustrates the test of hypothesis procedure to compare more than two independent means using ANOVA.

EXAMPLE 2-18

Hypertension, or high blood pressure, is the number one risk factor for death worldwide.[26] Eating a diet low in sodium, maintaining a healthy weight, not smoking, getting regular physical activity, and limiting consumption of alcoholic beverages have all been shown to reduce the risk of developing hypertension.[27] A study is conducted to evaluate the impact of physical activity on the progression to hypertension among normotensive[t] patients aged between 50 and 65 years. A total of 60 participants are randomly assigned to different levels of physical activity, classified as low, moderate, or high based on intensity and engagement in specific physical activities. Participants are supported and monitored to maintain the assigned level of physical activity for 1 year, at which time SBP and DBP are again measured. The primary outcome for this study is SBP and levels measured at 1 year are summarized in **TABLE 2-19**. Note that several participants did not complete the 1-year assessment, as the sample sizes in each group are less than 20 per group (as assigned). Loss to follow-up is an important issue that needs careful consideration in any analysis. Here, the loss is comparable across groups in terms of participant numbers, but it is always important to understand whether there are systematic issues that affect missingness as they could bias results.

(continues)

EXAMPLE 2-18 *(continued)*

TABLE 2-19 Systolic Blood Pressures by Physical Activity Level in Normotensive Patients Aged 50–65 Years

Physical Activity Level	Number of Participants	$\bar{X}(s)$ SBP at 1 Year (mmHg)
Low	17	135.7 (14.1)
Moderate	19	126.2 (12.7)
High	18	122.0 (13.1)

t Patients with systolic blood pressure (SBP), less than 120 mmHg, and diastolic blood pressure (DBP), less than 80 mmHg.

We wish to test whether there is a difference in mean SBP after 1 year following engagement in low, moderate, and high levels of physical activity among normotensive patients aged between 50 and 65 years.

1. Specify hypotheses and choose the level of significance.

    H_0: The mean SBPs in normotensive patients engaging in low, moderate, and high levels of physical activity for 1 year are equal ($\mu_1 = \mu_2 = \mu_3$).

    H_1: There is a difference in mean SBPs in normotensive patients engaging in low, moderate, and high levels of physical activity for 1 year (means not all equal[u]).

    $\alpha = 0.05$

    The null and research hypotheses could be stated differently as:

    H_0: There is no association between level of physical activity and mean SBP in normotensive patients.

    H_1: There is an association between level of physical activity and mean SBP in normotensive patients.

2. Compute test statistic based on sample data.

    The F statistic is computed as:

    $$F = \frac{\sum_{j=1}^{k} n_j \left(\bar{X}_j - \bar{X} \right)^2 \Big/ (k-1)}{\sum_{j=1}^{k} \left(n_j - 1 \right) s_j^2 \Big/ (N-k)},$$

    with an overall mean of 127.8.

---

u Note that we do not specify that $\mu_1 \neq \mu_2 \neq \mu_3$ in the research hypothesis as this is one of many possible alternatives to all means being equal. Rather we indicate in the research hypothesis that not all means are equal.

$$= \frac{\left[17(135.7-127.8)^2 + 19(126.2-127.8)^2 + 18(122.0-127.8)^2\right]/(3-1)}{\left(16\times14.1^2 + 18\times12.7^2 + 17\times13.1^2\right)/(54-3)} = 4.86.$$

The test statistic is $F = 4.86$.

3. Determine statistical significance of the data.

    The p-value is computed as $P(F > 4.86) = 0.012$.[v] At $\alpha = 0.05$, we reject the null hypothesis as the p-value ($p = 0.012$) is less than 0.05.

4. Summarize statistical and clinical or practical significance of the results.

    Participants assigned to the low physical activity group had a mean SBP of 135.7 mmHg at 1 year compared to 126.2 mmHg among those assigned to the moderate physical activity group and 122.0 mmHg among those assigned to the high physical activity group. This difference is statistically significant ($p = 0.012$).

    It is important to note that ANOVA tests the research hypothesis that the group means are not all equal. Here it seems that statistical significance is driven by the difference in SBP at 1 year between the low versus high physical activity groups. Those assigned to the moderate physical activity group have a mean SBP at 1 year that is reasonably close to that of the high physical activity group.

Multiple comparison procedures. ANOVA is a very useful hypothesis testing procedure for testing the equality of more than two independent means. When an ANOVA test is statistically significant, we conclude that the k means are not all equal. ANOVA does not convey which means are not equal. If the latter is of particular interest, we can conduct post hoc tests for specific differences in means. One particular type are pairwise tests, that is, tests of two means at a time. In an ANOVA with k comparison groups, there are $k(k - 1)/2$ possible pairwise tests. Each pairwise test could be conducted using the t-test approach outlined in the section on Hypothesis Tests Comparing Means in Two Independent Groups. However, each t-test has an associated Type I error probability (the probability that we reject the null hypothesis in favor of the research hypothesis when, in fact, the null hypothesis is true, e.g., $\alpha = 0.05$). When conducting multiple statistical tests, the overall Type I error probability can exceed what investigators consider acceptable. Multiple comparison procedures are statistical procedures that allow for multiple statistical tests of hypothesis to be conducted with control over the Type I error rate. Multiple comparison procedures are appropriate for post hoc tests following an ANOVA that reaches statistical significance.

---

v The f.dist.rt function in Excel computes p-values for ANOVA and requires three arguments: f.dist.rt(F, df$_1$, df$_2$). To compute the p-value with a test statistic of $F = 4.86$, df$_1 = 3 - 1 = 2$, and df$_2 = 54 - 3 = 51$, we specify f.dist.rt(4.86,2,51).

There are a number of popular multiple comparison procedures that differ based on the number and nature of the comparisons of interest, including Tukey, Dunnett, Scheffe, and Bonferroni procedures (see Cabral HJ[28] for more details). The Bonferroni procedure is the most conservative among these procedures, yet it is the easiest to apply and involves using a more stringent criterion for statistical significance for each post hoc test equal to the overall Type I error rate (e.g., $\alpha = 0.05$) divided by the number of post hoc tests. For example, for 3 post hoc tests, the significance criterion for each using Bonferroni's procedure is $0.05/3 = 0.0167$.

As an illustration, in Example 2-18 we observed a statistically significant difference among mean SBPs by physical activity groups ($p = 0.012$). There are three possible pairwise tests. For each test, a t statistic comparing the two independent means is computed and using the Bonferroni procedure we claim statistical significance for p-values that are less than 0.0167. The comparison of mean SBPs between those getting low versus moderate levels of physical activity produces $p = 0.041$, which would not be statistically significant. The comparison of mean SBPs between those getting moderate versus high levels of physical activity produces $p = 0.329$, which would also not be statistically significant. The comparison of mean SBPs between those getting low versus high levels of physical activity produces $p = 0.005$, which does reach statistical significance with the Bonferroni procedure.

Two-factor ANOVA. Analysis of variance is a widely used statistical approach for testing for differences among means. The procedure illustrated in Example 2-18 is called one-factor ANOVA as there is one grouping variable or factor. In some applications, there is more than one factor and ANOVA can be used to test for differences among means attributable to each of several factors. When there are exactly two grouping variables or factors, the analysis is called two-factor ANOVA. When there are three or more grouping variables or factors, the analysis is called higher-order ANOVA. **EXAMPLE 2-19** illustrates the approach when there are two factors. The same approach can be generalized to more than two factors, although the analysis becomes increasingly complex as there are many different possible relationships that might exist among multiple factors and the ways in which they influence the outcome of interest.

EXAMPLE 2-19

Consumption of highly processed foods has been shown to increase risks for cardiovascular and cerebrovascular disease.[29] Highly processed foods include sugar-sweetened beverages, sugary cereals, and ready-prepared meals that often contain added sugar, fat, and salt. A study is run to evaluate systolic blood pressure (SBP), a precursor to cardiovascular and cerebrovascular disease, in 1250 adults between the ages of 45 and 59 years according to the percentage

(continues)

EXAMPLE 2-19   (*continued*)

of their daily energy intake due to ultraprocessed foods and their age.
TABLE 2-20 summarizes SBP levels of participants by categories of percentage of their daily energy intake due to ultraprocessed foods and their age

TABLE 2-20 Mean (Standard Deviation) Systolic Blood Pressures by Categories of Percentage of Their Daily Energy Intake Due to Ultraprocessed Foods and Age

Percentage of Daily Intake of Energy Due to Ultraprocessed Foods (%)	45–49 Years of Age	50–54 Years of Age	55–59 Years of Age	Total
<10	127.3 (15.8) ($n = 100$)	129.4 (16.9) ($n = 125$)	135.3 (15.9) ($n = 150$)	131.2 (16.6) ($n = 375$)
10–20	134.2 (13.9) ($n = 125$)	138.2 (15.0) ($n = 150$)	142.9 (14.7) ($n = 160$)	138.8 (15.0) ($n = 435$)
>20	140.6 (15.7) ($n = 175$)	144.6 (15.8) ($n = 135$)	146.3 (14.0) ($n = 130$)	143.5 (15.4) ($n = 440$)
Total	135.3 (16.1) ($n = 400$)	137.6 (17.0) ($n = 410$)	141.3 (15.6) ($n = 440$)	138.2 (16.4) ($n = 1250$)

The mean SBP in the sample is 138.2 mmHg with a standard deviation of 16.4 mmHg. SBPs increase with age with mean SBPs of 135.3, 137.6, and 141.3 mmHg for people in the 45–49, 50–54, and 55–59 age groups, respectively. Similarly, SBPs increase with the percentage of daily intake of energy due to ultraprocessed foods with mean SBPs of 131.23, 138.8, and 143.5 mmHg for those <10%, 10%–20%, and >20% of daily intake of energy due to ultraprocessed foods, respectively. This is an example of a two-factor ANOVA and statistical tests are performed to assess the extent to which the differences in mean SBP are due to each factor. In this example, mean SBPs differ by both the percentage of daily intake of energy due to ultraprocessed foods ($p < 0.0001$) and by age group ($p < 0.0001$). **FIGURE 2-14** is an efficient way to display the data.

Note that Figure 2-14 shows the mean SBPs and standard error bars, computed as $s/\sqrt{n}$ for each group, which quantify sampling variability of each sample mean. The figure clearly shows a difference in SBP by age (within each category of daily energy intake due to ultraprocessed foods, SBP is higher among those in the 55–59-year-old age group compared to the 50–54 and 45–49 age groups) and also by percentage of daily energy

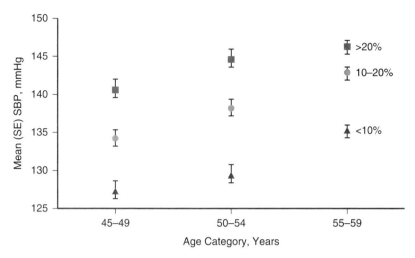

FIGURE 2-14 Mean (standard error) systolic blood pressures by categories of percent of their daily energy intake due to ultraprocessed foods and age.

intake due to ultraprocessed foods (within each age category, SBP is higher among those with >20% of daily energy intake due to ultraprocessed foods compared to 10%–20% and <10% of daily energy intake due to ultraprocessed foods).

Example 2-19 illustrates a two-factor ANOVA where each factor is statistically significant. Suppose the observed data were different and are summarized in **FIGURE 2-15**. In Figure 2-15, there is no statistical effect due to age. Within each category of daily energy intake due to ultraprocessed foods, the mean SBP are similar for each age group.

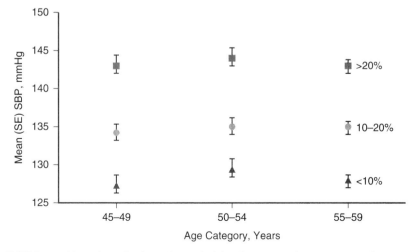

FIGURE 2-15 Mean (standard error) systolic blood pressures by categories of percent of their daily energy intake due to ultraprocessed foods and age: no effect due to age.

Suppose, yet again, that the observed data were different and are now summarized in **FIGURE 2-16**. In Figure 2-16, there is no statistical effect due to percentage of daily energy intake due to ultraprocessed foods. Within each age group, the mean SBPs are similar across categories of daily energy intake due to ultraprocessed foods.

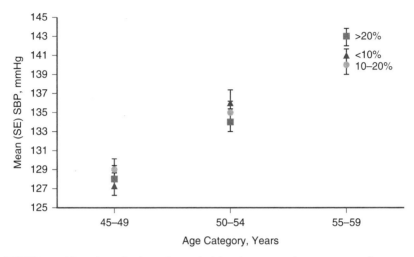

FIGURE 2-16 Mean (standard error) systolic blood pressures by categories of percent of their daily energy intake due to ultraprocessed foods and age: no effect due to percent daily energy intake due to ultraprocessed foods.

Consider Example 2-18 again, where we tested for differences in mean SBP by level of physical activity. Suppose that the investigators had randomized men and women separately to each of the physical activity groups and then wished to compare mean SBPs by physical activity level and gender. This is an example of a two-factor ANOVA with physical activity and gender as the two factors. Using two-factor ANOVA, we can test whether there is a difference in mean SBPs due to physical activity level, gender, or both. The latter term is called a statistical interaction. **TABLE 2-21** summarizes mean SBP levels measured at 1 year by physical activity group and gender.

TABLE 2-21 Mean Systolic Blood Pressures by Physical Activity Level and Gender in Normotensive Patients Aged 50–65 Years

Physical Activity Level	Men		Women	
	n	Mean SBP	*n*	Mean SBP
Low	8	141.1	9	130.8
Moderate	9	122.3	10	129.7
High	8	116.5	10	126.4

Note that among men, there appears to be a more pronounced impact of physical activity on SBP. In contrast, among women, the different levels of physical activity appear to have little impact on SBP. This is an example of a statistical interaction (also called effect modification by gender) where there is a different impact of one factor (physical activity) on the outcome (SBP) depending on the level of a second factor (gender), see **FIGURE 2-17**. Effect modification is discussed in more detail in Section 3.1 in the next unit.

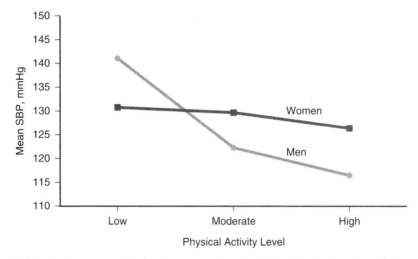

FIGURE 2-17 Mean systolic blood pressures by physical activity level and gender in normotensive patients, 50–65 years of age.

Statistical interactions can take many forms. In this example, there is a much stronger association between physical activity and SBP among men compared to women, as evidenced by the difference in mean SBPs in men with high versus moderate and low levels of physical activity. Among women, the differences in mean SBPs are practically negligible across the different categories of physical activity.

Repeated-measures ANOVA. In the section on Hypothesis Tests Comparing Means in Two Matched or Paired Groups we discussed tests of hypothesis for two matched or paired groups. An extension of this is called repeated-measures ANOVA, where there are more than two measurements per participant, often taken serially in time. When there are exactly two measurements, we take difference scores and test for a statistically significant mean difference using a *t*-test. When there are more than two measurements, we use a procedure called repeated measures ANOVA. In its simplest form, there is one within-subject factor where each participant is measured three or more times on a continuous outcome. The goal of the analysis is to evaluate the effect of time on mean outcome scores. This scenario can be extended to two-factor or higher-order ANOVA with repeated measures. The hypothesis testing approach is similar to that described above, but because repeated assessments are taken on the same participants, the test statistics must account for correlations within participants (see Sullivan LM[30] for more details).

Hypothesis Tests Comparing Proportions in Two or More Independent Groups

The *t*-tests to compare two independent and two matched or paired groups, and the ANOVA procedures to compare more than two groups are appropriate for continuous outcomes. In this section we describe a procedure, called the chi-square test, that is appropriate for comparing dichotomous, categorical, and ordinal outcomes among two or more independent groups.

In chi-square tests, the null hypothesis reflects the situation where the percentages of participants with the outcome, or the percentages in each outcome category, are equal across comparison groups. The research hypothesis reflects the situation where the percentages (or distribution of the outcome) is different among comparison groups. There is no direction of effect specified in the research hypothesis for chi-square tests. If any two comparison groups have different distributions of the outcome, then the research hypothesis is considered true.

The data for chi-square tests is best summarized in a cross-tabulation table. The table might be organized to show the comparison groups in the rows and the outcome of interest in the columns, or vice versa. The cell entries in the table represent the observed numbers of study participants in each comparison group by outcome category. The test statistic for chi-square tests is based on a comparison of these observed frequencies, or counts, to what would be expected if the null hypothesis were true. Specifically, chi-square test statistics are computed as: $\chi^2 = \dfrac{\Sigma(O - E)^2}{E}$, where O represents the observed and E the expected frequency in each cell of the cross-tabulation table. The test statistic compares observed and expected frequencies (by squaring the difference and dividing by the expected values) in each cell of the table which are then summed to produce the test statistic. The chi-square test is appropriate as long as the expected frequency in each cell is at least 5. If the expected frequency in any cell is less than 5, then Fisher's exact test is used (See[14] for more details). The chi-square statistic is always positive, and larger values suggest deviations from the null hypothesis and lead to statistical significance.

Statistical significance is summarized in a *p*-value; specifically the *p*-value is determined by $P(\chi^2 > \chi^2_{\text{test statistic}})$, where $\chi^2_{\text{test statistic}}$ is the observed test statistic. The χ^2 distribution is indexed by degrees of freedom, df, defined as the product of the number of groups -1 and the number of outcome categories -1. If the grouping variable is the row variable in the cross-tabulation table and r represents the number of rows/groups, and if the outcome variable is the column variable in the cross-tabulation table and c represents the number of columns/outcome categories, then the df $= (r-1)(c-1)$.

EXAMPLE 2-20 illustrates the chi-square test to compare outcome proportions in two independent groups.

EXAMPLE 2-20

Metabolic syndrome is a condition characterized by a cluster of risk factors including high levels of fat around the waist, elevated blood sugar level, systolic blood pressure, triglycerides, and cholesterol. It has been shown to increase the risk for Type II diabetes and cardiovascular disease.[31] A recent study reported that irregular sleep patterns were associated with increased risk for metabolic syndrome.[32] A survey of $n = 400$ adults between the ages of 40 and 75 years evaluated sleep patterns and prevalent metabolic syndrome in each participant. Sleep patterns were classified as regular or irregular, and participants with regular sleep patterns were those who went to bed and woke up at approximately the same times each day. The data are summarized in **TABLE 2-22**.

TABLE 2-22 Regular Sleep Patterns and Prevalent Metabolic Syndrome in Adults Aged 40–75 Years

	Metabolic Syndrome	No Metabolic Syndrome	Total
Regular sleep pattern	28	182	210
Irregular sleep pattern	47	143	190
Total	75	325	400

The prevalence of metabolic syndrome in the study sample is $75/400 = 18.8\%$. Among those with regular sleep patterns ($n = 210$), $28/210 = 13.3\%$ have metabolic syndrome compared to $47/190 = 24.7\%$ of those with irregular sleep patterns ($n = 190$). We wish to test whether the prevalence of metabolic syndrome in those with regular versus irregular sleep patterns are statistically different in patients between the ages of 40 and 75 years.

1. Specify hypotheses and choose the level of significance.

   H_0: The prevalence of metabolic syndrome in patients with regular sleep patterns is equal to the prevalence of metabolic syndrome in patients with irregular sleep patterns ($p_1 = p_2$).

   H_1: The prevalence of metabolic syndrome in patients with regular sleep patterns is different from the prevalence of metabolic syndrome in patients with irregular sleep patterns $(p_1 \neq p_2)$.

$\alpha = 0.05$

The null and research hypotheses could be stated differently as:

H_0: There is no association between sleep pattern and prevalent metabolic syndrome.

H_1: There is an association between sleep pattern and prevalent metabolic syndrome.

2. Compute test statistic based on sample data.

The χ^2 statistic is computed as: $\chi^2 = \dfrac{\Sigma(O - E)^2}{E}$, where O represents the observed frequencies that are reported in Table 2-22 and E represents expected frequencies, if the null hypothesis is true. To compute expected frequencies, we consider how many of the 210 and 190 patients with regular and irregular sleep patterns, respectively, we would have expected to have metabolic syndrome if the prevalence was equal in each group. To compute these expected frequencies, we apply the overall prevalence of 18.8% to each group total, as summarized in **TABLE 2-23**. Note that expected frequencies are computed to one decimal place as they are theoretical, as opposed to observed, frequencies.

TABLE 2-23 Expected Numbers of Patients With and Without Metabolic Syndrome, Assuming Prevalence of 18.8% in Each Group

	Metabolic Syndrome	No Metabolic Syndrome	Total
Regular sleep pattern	39.5	170.5	210
Irregular sleep pattern	35.7	154.3	190
Total	75	325	400

We now compute the chi-square test statistic by comparing observed and expected frequencies in each cell of the table:

$$\chi^2 = \frac{\Sigma(O - E)^2}{E} = \frac{(28 - 39.5)^2}{39.5} + \frac{(182 - 170.5)^2}{170.5}$$
$$+ \frac{(47 - 35.7)^2}{35.7} + \frac{(143 - 154.3)^2}{154.3}$$

$$= 3.35 + 0.78 + 3.58 + 0.83 = 8.53.$$

The test statistic is $\chi^2 = 8.53$.

3. Determine statistical significance of the data.

    The p-value is computed as $P(\chi^2 > 8.53) = 0.003$.[w] At $\alpha = 0.05$, we reject the null hypothesis as the p-value ($p = 0.003$) is less than 0.05.

4. Summarize statistical and clinical or practical significance of the results.

    Participants with regular sleep patterns have a prevalence of metabolic syndrome of 13.3% compared to 24.7% among those with irregular sleep patterns. This difference is statistically significant at $p = 0.003$.

EXAMPLE 2-21 illustrates the chi-square test to compare outcome proportions in three independent groups.

EXAMPLE 2-21

Consider, again, Example 2-19 where we analyzed differences in systolic blood pressure according to different percentages of daily energy intake due to ultraprocessed foods and age in a study sample of $n = 1250$ adults between the ages of 45 and 59 years. A subsample of $n = 600$ of these study participants agree to be followed for 15 years for the development of stroke. **TABLE 2-24** summarizes the numbers of patients in each category of percentage of daily energy intake due to ultraprocessed foods who suffer strokes by the end of the 15-year follow-up period.

TABLE 2-24 Number of Patients Who Develop a Stroke over 15 Years by Percent of Daily Energy Intake Due to Ultraprocessed Foods

Percentage of Daily Intake of Energy Due to Ultraprocessed Foods (%)	Stroke	No Stroke	Total
<10	6	189	195
10–20	9	191	205
>20	11	194	200
Total	26	574	600

We wish to test whether there is a difference in incident strokes by the percentage of daily intake of energy due to ultraprocessed foods.

---

w The chisq.dist.rt function in Excel computes the p-value for chi-square tests and requires two arguments: chisq.dist.rt(X^2, df). To compute the p-value with a test statistic of chi-square = 8.53, and df = $(2 - 1)(2 - 1) = 1$, we specify chisq.dist.rt(8.53, 1).

1. Specify hypotheses and choose the level of significance.

    H_0: The incidence of stroke in each category of percentage of daily intake of energy due to ultraprocessed food is equal $(p_1 = p_2 = p_3)$.

    H_1: The incidence of stroke in categories of percentage of daily intake of energy due to ultraprocessed foods are not all equal (proportions not all equal[x]).

    $\alpha = 0.05$

    The null and research hypotheses could be stated differently as:

    H_0: There is no association between percentage of daily intake of energy from ultraprocessed foods and incident stroke.

    H_1: There is an association between percentage of daily intake of energy from ultraprocessed foods and incident stroke.

2. Compute test statistic based on sample data.

    The χ^2 statistic is computed as: $\chi^2 = \dfrac{\Sigma(O-E)^2}{E}$, where

    O represents the observed frequencies that are reported in Table 2-24 and E represents expected frequencies, if the null hypothesis is true. To compute expected frequencies, we consider how many patients in each of the daily intake groups we would have expected to develop strokes over the 15-year follow-up period if the incidence was equal across groups. To compute these expected frequencies, we apply the overall incidence $26/600 = 0.043 = 4.3\%$ to each group total, as summarized in **TABLE 2-25**.

TABLE 2-25 Expected Numbers of Patients With and Without Incident Stroke, Assuming Incidence of 4.3% in Each Group

Percentage of Daily Intake of Energy Due to Ultraprocessed Foods (%)	Stroke	No Stroke	Total
<10	8.4	186.6	195
10–20	8.6	191.4	205
>20	8.8	196.2	200
Total	25.8	574.2	600

x Note that we do not specify that $p_1 \neq p_2 \neq p_3$ in the research hypothesis as this is one of many possible alternatives to all proportions being equal. Rather we indicate in the research hypothesis that not all proportions are equal.

Note that expected frequencies are computed to one decimal place as they are theoretical, as opposed to observed, frequencies.

We now compute the chi-square test statistic by comparing observed and expected frequencies in each cell of the table:

$$\chi^2 = \frac{\sum (O-E)^2}{E} = \frac{(6-8.4)^2}{8.4} + \frac{(189-186.6)^2}{186.6}$$
$$+ \frac{(9-8.6)^2}{8.6} + \frac{(191-191.4)^2}{191.4} + \frac{(11-8.8)^2}{8.8}$$
$$+ \frac{(194-196.2)^2}{196.2}$$
$$= 0.678 + 0.030 + 0.019 + 0.001 + 0.542 + 0.024$$
$$= 1.29$$

The test statistic is $\chi^2 = 1.29$.

3. Determine statistical significance of the data.

   The *p*-value is computed as $P(\chi^2 > 1.29) = 0.525$.[y] At $\alpha = 0.05$, we do not reject the null hypothesis as the *p*-value ($p = 0.525$) is not less than 0.05

4. Summarize statistical and clinical or practical significance of the results.

   Participants with <10% of their daily intake of energy due to ultraprocessed foods had 3.1% incidence of stroke over 15 years, compared to 4.5% among participants with between 10% and 20% of their daily intake of energy due to ultraprocessed foods, and 5.4% among participants with more than 20% of their daily intake of energy due to ultraprocessed foods. These differences, while potentially clinically meaningful, were not statistically significantly different ($p = 0.525$). This study may, in fact, be underpowered. The sample size ($n = 600$) is large, but there are relatively few outcome events (a total of 26 patients develop strokes). A larger group or a longer follow-up period might be required to detect statistical significance.

Hypothesis Tests Comparing Proportions in Two Matched or Paired Groups

In this section we discuss a specific chi-square test for the situation with a dichotomous outcome and two matched or paired groups. Specifically, we classify participants twice in terms of a dichotomous outcome (e.g., before and after an intervention). The null hypothesis is that proportions are equal in

---

y The chisq.dist.rt function in Excel computes *p*-values for chi-square tests and requires two arguments: chisq.dist.rt(χ^2, df). To compute the *p*-value with a test statistic of $\chi^2 = 1.29$ and df $= (3-1)(2-1) = 2$, we specify chisq.dist.rt(1.29,2).

the matched or paired groups and the research hypothesis reflects the situation where one proportion is higher than the other or that the proportions are different, whichever the investigator believes to be true.

The chi-square test for this specific application with a dichotomous outcome and two matched or paired groups is called McNemar's test. Again, the data are summarized in a cross-tabulation table with the rows and columns of the table showing each outcome state (e.g., disease, no disease) for the first and second assessment. The cell entries in the table represent the observed numbers of study participants in each combination of outcome states. **TABLE 2-26** illustrates the data layout for McNemar's test.

TABLE 2-26 Data Layout for McNemar's Test of Two Matched or Paired Proportions

	First Assessment – Positive	First Assessment – Negative	Total
Second assessment – positive	a	b	
Second assessment – negative	c	d	
			n

The proportion of positive responses as per the first assessment is $\hat{p}_1 = \dfrac{(a+c)}{n}$ and the proportion of positive responses as per the second assessment is $\hat{p}_2 = \dfrac{(a+b)}{n}$. The research question concerns the equality of population proportions, which is judged based on a comparison of observed sample proportions.

The test statistic for McNemar's test is a chi-square statistic, computed as: $\chi^2 = \dfrac{(b-c)^2}{(b+c)}$, where b and c represent the numbers of discordant pairs in Table 2-26 (i.e., the total number of participants classified as negative on the first assessment and positive on the second, positive on the first assessment and negative on the second, respectively). This chi-square statistic is always positive, and larger values suggest deviations from the null hypothesis and lead to statistical significance.

Statistical significance is summarized in a p-value, specifically the p-value is determined by $P(\chi^2 > \chi^2_{\text{test statistic}})$, where $\chi^2_{\text{test statistic}}$ is the observed test statistic. The χ^2 distribution is indexed by degrees of freedom and for McNemar's test with a dichotomous outcome and two matched or paired groups, df $= 1$.

EXAMPLE 2-22 illustrates McNemar's test to compare outcome proportions in two matched groups.

EXAMPLE 2-22

Attention-deficit/hyperactivity disorder (ADHD) is a highly prevalent condition characterized by frequent episodes of inattention, impulsivity, and hyperactivity that impacts those who suffer from ADHD personally, socially, at school, and at work. ADHD is a treatable medical condition when diagnosed correctly. An investigator proposes a new holistic treatment for patients newly diagnosed with ADHD. The treatment is specifically designed to address impulsivity. A total of $n = 320$ patients with ADHD participate in the study and each patient is asked to rate their symptoms of impulsivity on an extensive checklist before and after receiving the new treatment. The symptom checklists are reviewed by psychologists and each patient is classified as having, or not having, impulsivity issues before and after treatment. The data are summarized in **TABLE 2-27**.

TABLE 2-27 Impulsivity Before and After Receiving Treatment for ADHD

After Treatment	Before Treatment		
	Impulsivity Present	Impulsivity Absent	
Impulsivity present	100	15	115
Impulsivity absent	52	153	
	152		320

Before treatment, $152/320 = 0.475 = 47.5\%$ of patients had symptoms of impulsivity. After treatment, $115/320 = 0.359 = 35.9\%$ of patients had symptoms of impulsivity. We wish to test whether there is a statistically significant difference in the proportion of patients with symptoms of impulsivity before versus after treatment in patients with ADHD.

1. Specify hypotheses and choose the level of significance.

   H_0: The proportion of patients with impulsivity before treatment is equal to the proportion of patients with impulsivity after treatment ($p_1 = p_2$).

   H_1: The proportion of patients with impulsivity before treatment is not equal to the proportion of patients with impulsivity after treatment ($p_1 \neq p_2$).

   $\alpha = 0.05$

2. Compute test statistic based on sample data.

   Because we have two matched groups, we use McNemar's test and compute the χ^2 statistic as: $\chi^2 = \dfrac{(b-c)^2}{(b+c)}$, where b and c

represent the number of patients with symptoms of impulsivity after treatment and not before, and the number of patients with symptoms of impulsivity before treatment and not after treatment, respectively.

$$\chi^2 = \frac{(b-c)^2}{(b+c)} = \frac{(15-52)^2}{15+52} = 20.43.$$

The test statistic is $\chi^2 = 20.43$.

3. Determine statistical significance of the data.

The p-value is computed as $P(\chi^2 > 20.43) < 0.0001$.[z] At $\alpha = 0.05$, we reject the null hypothesis as the p-value ($p < 0.0001$) is less than 0.05.

4. Summarize statistical and clinical or practical significance of the results.

Before treatment, 47.5% of patients had symptoms of impulsivity compared to 35.9% of patients having symptoms of impulsivity after treatment. This difference is statistically significant at $p < 0.0001$.

▶ 2.6 Correlation Analysis

In some investigations we wish to understand whether, and to what extent, two continuous or measurement variables might be related (e.g., age and weight, caffeine intake and blood sugar level, systolic blood pressure and body mass index, measures of mental health and physical health). The Pearson's Product moment correlation coefficient, r, quantifies the direction (positive or negative) and strength of the linear association between two continuous or measurement variables. The correlation coefficient ranges from −1 to 1 and the sign indicates the direction of the association and the magnitude indicates the strength of the association. A correlation of 0 indicates no linear association between the two variables. A correlation of 0 could mean that there truly is no relationship between the variables or that a nonlinear association exists. Thus, it is good practice to generate a scatter plot as the first step in any correlation analysis (more on this in the examples that follow).

Estimating a Correlation

With correlation analysis, it may be that the variables play specific roles (i.e., one is a predictor and the other an outcome), or do not play specific roles (e.g., measures of mental health and physical health). The computation of

---

z   The chisq.dist.rt function in Excel computes p-values for chi-square tests and requires two arguments: chisq.dist.rt(χ^2, df). To compute the p-value with a test statistic of $\chi^2 = 20.43$ and df = 1, we specify chisq.dist.rt(20.43, 1).

the correlation coefficient is the same for each scenario. To simplify computational formulas, we designate one of the variables, x, and the other, y. If one variable plays the role of predictor, it is labelled x, and the outcome is y. If not, we simply assign the x and y labels to the two variables. The correlation coefficient is computed as: $r = \dfrac{\Sigma(x - \bar{x})(y - \bar{y})}{\sqrt{s_x^2 s_y^2}}$, where $s_X^2 = \dfrac{\Sigma(x - \bar{x})^2}{n - 1}$

is the sample variance of x, and $s_y^2 = \dfrac{\Sigma(y - \bar{y})^2}{n - 1}$ is the sample variance of y. If the correlation, r, is close to 0, the correlation between the two variables is described as very weak. Some consider correlations between $|0{-}0.2|$ as very weak, correlations between $|0.2{-}0.4|$ as weak, $|0.4{-}0.6|$ as moderate, $|0.6{-}0.8|$ as strong, and $|0.8{-}1.0|$ as very strong. These are general guidelines for interpretation. It is also possible to run a test of hypothesis to determine whether the true population correlation, ρ, is statistically significantly different from 0 (the null value). **EXAMPLES 2-23** and **2-24** illustrate this approach.

EXAMPLE 2-23

We wish to examine the correlation between BMI and high-density lipoprotein (HDL) cholesterol in a large cross-sectional study of $n = 5{,}413$ adults between the ages of 50 and 75 years. Both variables are continuous with BMI measured in kg/m² and HDL measured in mg/dL. HDL cholesterol is sometimes called the "good" cholesterol and higher levels are better than lower levels. In the study sample, the mean (standard deviation) of BMI is 26.40 (4.28) and the mean (standard deviation) of HDL is 51.67 (15.86). **FIGURE 2-18** is a scatter plot showing the relationship between BMI and HDL in the study sample. In this example, we consider BMI as the predictor (shown on the x-axis) and HDL cholesterol as the outcome (shown on the y-axis).

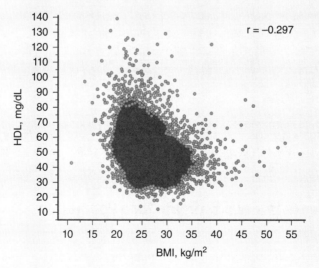

FIGURE 2-18 Scatter plot of the association between BMI and HDL cholesterol.

We use a statistical computing package to estimate the correlation coefficient and the result is $r = -0.297$, which suggests that there is a weak, negative, or inverse association between BMI and HDL cholesterol.

EXAMPLE 2-24

We wish to examine the correlation between gestational age, measured in weeks, and birthweight, measured in grams in $n = 486$ women with a history of premature birth. In this example, we consider gestational age as the predictor, x, and birthweight as the outcome, y. In the study sample, the mean (standard deviation) of gestational age is 33.31 (3.98) weeks and the mean (standard deviation) of birthweight is 2,071.62 (858.45) g. **FIGURE 2-19** is a scatter plot showing the relationship between gestational age and birthweight in the study sample.

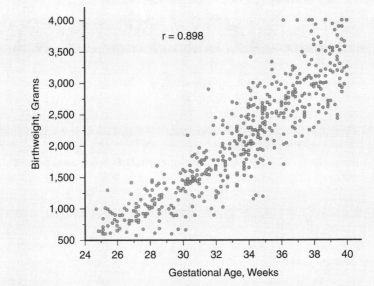

FIGURE 2-19 Scatter plot of the association between gestational age and birthweight.

We use a statistical computing package to estimate the correlation coefficient and the result is $r = 0.898$, which suggests that there is a very strong, positive, or direct association between gestational age and birthweight.

Confidence Interval Estimate for a Population Correlation

The sample correlation, r, is a point estimate for the true, population proportion. In order to generate a CI estimate for the true population proportion,

ρ, we need to transform the observed correlation, r, using Fisher's z transformation which promotes normality and allows us to use the standard normal distribution approach outlined for means and proportions to compute the CI. Specifically, we first transform the observed correlation, r, using Fisher's z transformation, as: $z_r = 0.5\ln\dfrac{(1+r)}{(1-r)}$. Second, we estimate the CI for z_r using: $\left(z_r - z\dfrac{1}{\sqrt{n-3}}, z_r + z\dfrac{1}{\sqrt{n-3}}\right)$, where z is the value from the standard normal distribution reflecting the desired confidence level (e.g., for 95%, $z = 1.96$). And, finally, we back transform the lower and upper limits of the CI for z_r to produce the CI for r using: $\text{lower}_r = \dfrac{e^{2(\text{lower } z_r)}-1}{e^{2(\text{lower } z_r)}+1}$ and: $\text{upper}_r = \dfrac{e^{2(\text{upper } z_r)}-1}{e^{2(\text{upper } z_r)}+1}$. We illustrate these steps in the following examples.

In Example 2-23, we estimated the correlation between BMI and HDL cholesterol as $r = -0.297$ and concluded that there was a weak, negative, or inverse association between BMI and HDL cholesterol. We now estimate a 95% CI for the true correlation between BMI and HDL cholesterol using Fisher's transformation. First, we transform the observed correlation, r, using Fisher's z transformation, as: $z_r = 0.5\ln\dfrac{(1+(-0.297))}{(1-(-0.297))} = -0.306$. Second, we estimate a CI for z_r using: $\left(z_r - z\dfrac{1}{\sqrt{n-3}}, z_r + z\dfrac{1}{\sqrt{n-3}}\right)$. Substituting $z_r = -3.06$, $z = 1.96$ for 95% confidence, and $n = 5{,}413$, we get $\left(-0.306 - 1.96\dfrac{1}{\sqrt{5{,}413-3}}, -0.306 + 1.96\dfrac{1}{\sqrt{5{,}413-3}}\right)$, which is equivalent to $(-0.333, -0.279)$ and reflects a 95% CI for z_r. The final step involves back transforming the lower and upper limits of the CI for z_r to produce the CI for r, as: $\text{lower}_r = \dfrac{e^{2(-0.333)}-1}{e^{2(-0.333)}+1}$ and: $\text{upper}_r = \dfrac{e^{2(-0.279)}-1}{e^{2(-0.279)}+1}$, which is equivalent to $(-0.321, -0.272)$ and reflects a 95% CI for r. Because the 95% CI does not include 0 (the null value), we conclude that the true population correlation between BMI and HDL cholesterol is statistically significantly different from 0.

In Example 2-24, we estimated the correlation between gestational age and birthweight $r = 0.898$ and concluded that there was a very strong, positive, or direct association between gestational age and birthweight. Most statistical computing packages produce CI estimates for correlations. The 95% CI estimate for the correlation between gestational age and birthweight is $(0.879, 0.914)$. Because the 95% CI does not include 0 (the null value), we conclude that the true population correlation between gestational age and birthweight is statistically significantly different from 0.

Hypothesis Test for a Population Correlation

In some applications, it might be of interest to test whether there is a statistically significant correlation in the population based on the observed sample correlation. This is equivalent to testing whether the true population correlation, ρ, is significantly different from 0. The null hypothesis reflects the situation where there is no correlation in the population (i.e., $\rho = 0$) and the research hypothesis specifies that there is a correlation in the population (i.e., $\rho \neq 0$). The test statistic is: $t = r\sqrt{\dfrac{(n-2)}{(1-r^2)}}$, which has df =

$n - 2$, and p-values are calculated by computing probabilities of observing a test statistic as, or more extreme than, that observed using the t probability distribution.

In Example 2-23, we estimated the correlation between BMI and HDL cholesterol as $r = -0.297$. We could conduct a test of hypothesis to evaluate whether this is significant evidence of a correlation in the population. The

test statistic is: $t = -0.297\sqrt{\dfrac{(5413-2)}{\left[1-\left(-0.297\right)^2\right]}} = -22.879$, which is highly statis-

tically significant at $p < 0.0001$.

Thus, we conclude that the true population correlation between BMI and HDL cholesterol is statistically significantly different from 0, which is consistent with the interpretation based on the CI estimate.

In Example 2-24, we estimated the correlation between gestational age and birthweight $r = 0.898$. We compute a significance test using a statistical computing package and observe $p < 0.0001$. Thus, we conclude that the true population correlation between gestational age and birthweight is statistically significantly different from 0, which is consistent with the interpretation based on the CI estimate.

▶ 2.7 Summary

In many applications we are interested in the association between an exposure or risk factor and a health outcome in a population. We select a sample from that population that is ideally representative of the population, and analyze the association in the sample. We then make inferences about the true association based on what is observed in the sample, using one of two popular approaches, estimation or hypothesis testing. Estimation involves generating estimates of unknown population parameters including means, proportions, and comparisons of means and proportions using differences or ratios in independent and matched or paired groups based on data in the study sample. Hypothesis testing involves setting up competing hypotheses, the null and research hypotheses, which represent no association and association in a specific direction, or not, respectively, and analyzing sample data to infer whether the research hypothesis is likely true.

The theoretical underpinning of estimation and hypothesis testing is the central limit theorem that states that the distribution of sample means is approximately normal regardless of the distribution of the outcome in the population, as long as the sample is sufficiently large. This is an important result that enables us to use the properties of the normal distribution in estimation and hypothesis testing.

In this unit, we outlined different formulas for CI estimates that depend on the nature of the outcome (continuous or dichotomous), the number of comparison groups and the relationship among those groups (i.e., whether they are independent, matched, or paired). Each CI takes the same form: point estimate ± margin of error, and the resultant interval is the range of plausible values for the unknown population parameter.

We also discussed hypothesis testing procedures following the same general approach in each case. The specific test statistic, which summarizes the sample data, depends on the nature of the outcome (continuous or dichotomous), the number of comparison groups, and the relationship among those groups (i.e., whether they are independent, matched, or paired). In hypothesis tests for one and two means, we use t-tests, and for comparing more than two means we use analysis of variance (ANOVA). Comparisons of two or more proportions are based on chi-square tests.

We now return to the population health issue we started with—future health risks associated with physical inactivity among children—and apply some of the techniques we discussed in this unit to address each question.

- How likely are children worldwide to meet the recommended physical activity guidelines?

  The World Health Organization provides facts on physical activity for children and adults worldwide and reports that 81% of adolescents between the ages of 11 and 17 years were not meeting physical activity guidelines in 2010, with adolescent girls (84%) being more likely to not meet physical activity goals compared to boys (78%) meeting these requirements.[33]

  A recent study of nearly 18,000 undergraduate students from low-, middle-, and high-income countries showed wide variation in prevalence of physical inactivity among male and female students.[34] For example, in Caribbean countries, 28.7% of male students compared to 42.7% of female students were physically inactive ($p < 0.001$). In Sub-Saharan Africa, 30.9% of male students compared to 41.3% of female students were physically inactive ($p < 0.001$). In the Near East and Central Asia, 31.8% of male students compared to 32.9% of female students were physically inactive ($p < 0.340$). And in South Asia, China, and Southeast Asia, about 37% of male students and 52% of female students were physically inactive ($p < 0.001$).

- What are the health benefits and consequences of physical inactivity?

  Regular physical activity improves muscle mass, strengthens bones and joints, improves cardiorespiratory fitness, reduces levels of stress and anxiety, promotes psychosocial health, and helps in maintaining a healthy weight. Physical inactivity is increasing in many countries and is one of the leading risk factors for death worldwide. People who are less physically active are at higher risk for obesity, high blood pressure, Type II diabetes, anxiety and depression, cardiovascular disease, and certain cancers later in life.[35] For example, a study of nearly 25,000 adults between the ages of 39 and 79 years showed that active adults had about half the risk—0.54, with a 95% CI (0.50–0.59)—of developing cardiovascular disease compared to their inactive counterparts.[36]

- What interventions might be effective in increasing physical activity in children?

  Healthy People 2020 is a set of goals based on scientific data to improve the health of all Americans, and includes a specific goal around improving health and quality of life through increased physical activity.[37] Some structural changes are recommended that might promote increases in physical activity, including improvements to the built environment such as sidewalks, bike lanes, walking trails, and parks, as are new policies to improve access to facilities that promote physical activity. Behavioral interventions are also recommended to reduce television watching and computer time. Each of these interventions, as well as others, could be tested for their impact on physical activity using the tests of hypothesis discussed in this unit.

The hypothesis testing procedures outlined here are widely used in practice and are generally described as superiority tests where the goal is to produce statistical support in favor of the research hypothesis. There are other applications where it is of interest to provide evidence in support of a null hypothesis. Such tests are called equivalence, or noninferiority, tests and follow a different approach than the four-step approach outlined in Section 2.5.[38,39] Such tests are used when it is of interest to show that a new test or treatment is as effective as a comparison test or treatment, and would be justified based on other attributes such as lower cost or fewer side effects.

Also, the hypothesis testing procedures outlined here for continuous outcomes are parametric tests that make assumptions about the distributional form of the outcome in the population. For example, if a continuous outcome is reasonably assumed to follow a normal distribution, then a test of hypothesis involving a t statistic is justified. However, when the distribution of a continuous outcome is not known and the sample size is small, non-parametric tests might be more appropriate. Parametric tests for continuous outcomes focus on means, whereas nonparametric tests for continuous variables focus on medians. Thus, if a median is a better representation of a typical value for a particular outcome then a nonparametric test might be preferable. It is important to note that parametric tests are more powerful than nonparametric tests (i.e., if the research hypothesis is true, a parametric test is more likely to detect it). Some popular nonparametric alternatives to the tests we discussed in this unit are the one-sample sign test for a median or the Wilcoxon signed rank test for a median as alternatives for the one sample t-test for a mean; the Mann-Whitney U test for equality of medians as an alternative to the two independent samples t-test for equality of means; the one-sample sign test or the Wilcoxon signed rank test for a median as alternatives to the two sample t-test for the mean difference in matched or paired samples; and the Kruskal-Wallis test for equality of medians as an alternative to one-factor ANOVA. (More details on these and other nonparametric tests can be found in Corder and Foreman[14] and Sullivan LM[40.])

Key Points

- Probabilities are computed by dividing the number of participants with a particular attribute or outcome by the total number in the population. In some applications, we condition, or focus, on a particular subset of the population, but follow the same approach in computing probabilities.
- Screening tests are used for early detection of disease. Performance characteristics, including sensitivity, specificity, false-positive and false-negative fractions, summarize the accuracy of screening tests and inform patients as to whether they should take a particular test. Positive and negative predictive values are important in interpreting screening test results. Positive and negative predictive values are influenced by the prevalence of the disease or condition under study.
- The normal distribution is a popular probability distribution model for continuous outcomes. The normal distribution is completely defined by its mean and standard deviation. An outcome that follows a normal distribution has specific properties, including the mean = median = mode, and that approximately 95% of values fall between the mean plus 2 or minus 2 standard deviations.
- The standard normal distribution, z, is a specific normal distribution with a mean of 0 and a standard deviation of 1, and is used in many statistical applications.
- Confidence interval (CI) estimates for any unknown population parameter take the form: point estimate ± margin of error.
- If a CI includes the null value of a population parameter (e.g., the null value of a difference in means is 0, the null value of a relative risk or risk ratio is 1), then there is no statistical evidence of a difference or an association.

- Nonoverlapping CIs for independent groups suggest a statistical difference between groups, but overlapping CIs do not ensure that there is no statistical difference. Thus, if interest lies in comparing independent groups, it is appropriate to compute a CI for the difference or ratio directly rather than comparing CIs for each group, considered separately.
- It is very important to differentiate applications with two independent groups versus matched or paired groups. Applying statistical procedures for independent groups when the groups are matched or paired can lead to incorrect conclusions.
- p-values summarize the incompatibility of sample data with the assumed statistical model. A smaller p-value (e.g., $p < 0.05$) suggests statistical evidence in favor of the research hypothesis.
- p-values summarize statistical significance. Clinical or practical significance is judged by the investigator based on content expertise.

References

1. World Health Organization. Global Health Observatory (GHO) data. https://www.who.int/gho/ncd/risk_factors/en/. Accessed March 3, 2019.
2. Active Healthy Kids Global Alliance. https://www.activehealthykids.org/2018/11/26/childhood-physical-inactivity-reaches-crisis-levels-around-the-globe/. Accessed March 3, 2019.
3. *National Physical Actvity Plan Alliance. The 2018 United States Report Card on Physical Activity for Children and Youth.* Washington, DC: National Physical Activity Plan Alliance; 2018.
4. Baio J, Wiggins L, Christensen DL, et al. Prevalence of autism spectrum disorder among children aged 8 years—autism and developmental disabilities monitoring network, 11 sites, United States, 2014. *MMWR Surveill Summ.* 2018;67(No. SS-6):1-23.
5. TDR. Disease watch focus: syphilis. https://www.who.int/tdr/publications/disease_watch/syphilis/en/. Accessed April 4, 2019.
6. Centers for Disease Control and Prevention. Sexually transmitted disease surveillance 2017: syphilis. https://www.cdc.gov/std/stats17/syphilis.htm. Accessed April 4, 2019.
7. Jafari Y, Peeling RW, Shivkumar S, Claessens C, Joseph L, Pai NP. Are Treponema pallidum specific rapid and point-of-care tests for syphilis accurate enough for screening in resource limited settings? Evidence from a meta-analysis. *PLoS One* 2013;8(2):e54695.
8. Nah E-H, Cho S, Kim S, Cho H-I, Chai J-Y. Comparison of traditional and reverse syphilis screening algorithms in medical checkups. *Ann Lab Med.* 2017;37:511-515.
9. Rubio-Tapia A, Ludvigsson JF, Brantner TL, Murray JA, Everhart JE. The prevalence of celiac disease in the United States. *Am J Gastroenterol.* 2012;107(10):1538-1544.
10. National Institutes of Diabetes and Digestive and Kidney Diseases. Celiac disease testing (for healthcare professionals). https://www.niddk.nih.gov/health-information/diagnostic-tests/celiac-disease-health-care-professionals. Accessed March 3, 2019.
11. Framingham Heart Study. History of the Framingham Heart Study. https://www.framinghamheartstudy.org/fhs-about/history/. Accessed April 2, 2019.

12. National Heart, Lung, and Blood Institute. Teaching datasets. https://biolincc.nhlbi .nih.gov/teaching/. Accessed April 16, 2019.

13. Centers for Disease Control and Prevention. Clinical growth charts. https://www .cdc.gov/growthcharts/clinical_charts.htm. Accessed June 15, 2019.

14. Corder GW, Foreman DI. *Nonparametric Statistics. A Step-by-Step Approach*. 2nd ed. Hoboken, NJ: John Wiley and Sons, Inc; 2014.

15. Centers for Disease Control and Prevention. Leading causes of death. https://www .cdc.gov/nchs/fastats/leading-causes-of-death.htm. Accessed April 27, 2019.

16. Centers for Disease Control and Prevention. Heart disease facts. https://www.cdc .gov/heartdisease/facts.htm. Accessed April 27, 2019.

17. Koiinis-Mitchell D, Rosario-Matos N, Ramirez RR, Garcia P, Canino GJ, Ortega AN. Sleep, depressive/anxiety disorders, and obesity in Puerto Rican youth. *J Clin Psychol Med Settings* 2017;24(1):59-73.

18. Aschengrau A, Seage GR III. *Essentials of Epidemiology in Public Health*. 2nd ed. Sudbury, MA: Jones and Bartlett Publishers; 2008.

19. Fagerland MW, Lyderson S, Laake P. Recommended tests and confidence intervals for paired binomial proportions. *Stat Med*. 2014;33:2850-2875.

20. Fleiss JL, Levin B, Paik MC. *Statistical Methods for Rates and Proportions*. 3rd ed. New York, NY: John Wiley & Sons, Inc.; 2003.

21. Wasserstein RL, and On Behalf of the American Statistical Association Board of Directors. The ASA's statement on p-values: context, process, and purpose. *Am Stat*. 2016;70(2):129-133.

22. Centers for Disease Control and Prevention. Get the facts: sugar-sweetened beverages and consumption. https://www.cdc.gov/nutrition/data-statistics/sugar -sweetened-beverages-intake.html. Accessed May 26, 2019.

23. *Dietary Guidelines for Americans 2015–2020*. 8th ed. https://health.gov /dietaryguidelines/2015/guidelines/?linkId=20169028. Accessed May 26, 2019.

24. Wing RR, Phelan S. Long-term weight loss maintenance. *Am J Clin Nutr*. 2005;82(1):222S-225S.

25. E-cigarette use among youth and young adults. A report of the surgeon general. https://www.cdc.gov/tobacco/data_statistics/sgr/e-cigarettes/pdfs/2016_sgr _entire_report_508.pdf. Accessed May 27, 2019.

26. Our World in Data. https://ourworldindata.org/grapher/number-of-deaths-by-risk -factor. Accessed May 29, 2019.

27. Centers for Disease Control and Prevention. Preventing high blood pressure: healthy living habits. https://www.cdc.gov/bloodpressure/healthy_living.htm. Accessed May 29, 2019.

28. Cabral HJ. Multiple comparison procedures. *Circulation* 2008;117:698-705.

29. Srour B, Fezeu LK, Kesse-Guyot E, et al. Ultra-processed food intake and risk of cardiovascular disease; prospective cohort study (NutriNet-Sante). *BMJ* 2019;365:I1451.

30. Sullivan LM. Repeated measures. *Circulation* 2008;117:1238-1243.

31. Wilson PW, D'Agostino RB, Parise H, Sullivan LM, Meigs JB. Metabolic syndrome as a precursor of cardiovascular disease and type 2 diabetes mellitus. *Circulation* 2005;112(20):3066-3072.

32. Huang T, Redline S. Cross-sectional and prospective associations of actigraphy-assessed sleep regularity with metabolic abnormalities: the Multi-Ethnic Study of Atherosclerosis. *Diabetes Care* 2019;42(8):1422-1429.

33. World Health Organization. Physical activity fact sheet. https://www.who.int/news -room/fact-sheets/detail/physical-activity. Accessed June 11, 2019.

34. Pengpi S, Peltzer K, Kassean HK, Tsala JPT, Sychareun V, Muller-Riemenschneider F. Physical inactivity and associated factors among university stuents in 23 low-, middle- and high-income countries. *Int J Public Health* 2015;60:539-549.
35. Gonzalez K, Fuentes J, Marquez JL. Physical inactivity, sedentary behavior and chronic diseases. *Korean J Fam Med.* 2017;38(3):111-115.
36. Lachman S, Boekholdt SM, Luben RN, et al. Impact of physical activity on the risk of cardiovascular disease in middle-aged and older adults; EPIC Norfolk prospective population study. *Eur J Prev Cardiol.* 2018;25(2):200-208.
37. *Healthy People 2020.* About healthy people. https://www.healthypeople.gov/2020/About-Healthy-People. Accessed June 13, 2019.
38. D'Agostino RB Sr, Massaro J, Sullivan LM. Non-inferiority trials: design concepts and issues - the encounters of academic consultants in statistics. *Stat Med.* 2003;22(2):169-186.
39. Tunes da Silva GT, Logan BR, Klein JP. Methods for equivalence and noninferiority testing. *Biol Blood Marrow Transplant.* 2008;15(1 Suppl.):120-127.
40. Sullivan LM. *Essentials of Biostatistics in Public Health.* 3rd ed. Burlington, MA: Jones & Bartlett Learning; 2018.

UNIT 3
Multivariable Analysis

In Unit 2, we outlined estimation and hypothesis testing techniques to evaluate the association between an exposure or risk factor and a health outcome. We selected a sample from the population of interest, analyzed the association in the sample and made inferences about the true association in the population based on what was observed in the sample. In each application, there was one exposure or risk factor (e.g., assigned intervention or treatment, a personal attribute or a specific behavior) and one health outcome (e.g., disease severity, prevalent or incident disease).

In this unit, we discuss multivariable analysis, moving beyond the association between two variables to take into account relationships among other variables that are inevitably at play. We first define bias, confounding and effect modification. We then discuss techniques to address confounding in particular, in order to isolate, insofar as possible, the impact of a key risk factor, exposure, or predictor, on the outcome of interest taking into account other variables.

In many applications, the goal is to evaluate the association between a primary or key exposure or risk factor and an outcome, while accounting for other risk factors that may play a role. In other applications, the goal is to evaluate the relative importance of a number of risk factors or exposures on a health outcome. We describe and illustrate popular multivariable statistical applications such as multivariable linear regression analysis, multivariable logistic regression analysis, and survival analysis that can be used to address these goals. With each technique, we estimate a regression equation that relates a set of risk factors or predictors to the outcome of interest. We test whether there is a statistically significant association between the set of risk factors and the outcome, and then evaluate how each risk factor is related to the outcome in the presence of the others.

A Population Health Issue—Diabetes Prevalence Increasing Worldwide

Diabetes is a condition where the body either does not produce enough insulin (Type I diabetes) or cannot use the insulin that is produced (Type II diabetes), with the latter being far more prevalent. Insulin is important as it controls blood sugar levels. A third classification is gestational diabetes that occurs during pregnancy among women who had normal blood sugar levels prior to becoming pregnant. People with Type I and Type II diabetes are at higher risk for cardiovascular disease, nerve damage in the feet, vision problems, and kidney failure.[1] Women with gestational diabetes are at higher risk for complications of pregnancy, adverse outcomes of pregnancy such as premature delivery, and their children are at elevated risk for obesity and for developing Type II diabetes later in life.[2] In fact, a recent study of over 70,000 pregnancies in Quebec, Canada, found that children born to mothers with gestational diabetes had almost twice the risk of developing diabetes before reaching the age of 22.[3]

The World Health Organization (WHO) reports that as of 2014, there were 422 million people living with diabetes worldwide and that number is rising, particularly in middle- and low-income countries.[4] The WHO also reports that diabetes was the seventh leading cause of death worldwide in 2016, with over 1.5 million deaths attributable to diabetes.

In this unit, we discuss biostatistical techniques that will allow us to answer questions such as the following:

- What are the modifiable risk factors for Type I and Type II diabetes?
- Are there effective treatments for gestational diabetes to minimize the risks of adverse pregnancy outcomes?
- Are patients with Type II diabetes at higher risk of death due to cardiovascular disease?

Before we delve into the details of how to construct multivariable models to answer these and other important questions, we first define bias, confounding, and effect modification; describe approaches to address confounding; and then outline the data considerations for multivariable models.

▶ 3.1 Bias, Confounding, and Effect Modification

Bias is a systematic error in study design, recruitment of participants, measurement, or statistical analysis that results in an incorrect assessment of the true association between a risk factor or predictor and an outcome. There are many different sources of bias that can be generally classified as selection bias or information bias. Selection bias occurs when there are systematic errors in the selection of participants or retention of participants over time in a study sample. Selection bias results in a sample that is no longer representative of the population, leading to incorrect inferences. Information bias occurs when there are systematic errors in measurement of risk factors or outcomes, or in the classification of participants in terms of risk factor, exposure or outcome status. Again, the result is that the sample is no longer representative of the population, leading to incorrect inferences. Bias is caused by the investigator in the design, conduct, or analysis of the study and cannot be corrected. Thus, the best approach to handling bias involves taking steps to avoid it.

Confounding is a distortion, exaggeration, or masking of an association between a risk factor and an outcome due to other variables. As most health outcomes are affected by multiple risk factors, as opposed to a single risk factor, confounding is an issue in many health-related applications. There are three conditions that define a confounder. First, the confounder is associated with the outcome of interest. Second, the confounder is associated with the risk factor of interest. And third, the confounder is not in the causal pathway between the risk factor and the outcome. Positive confounding occurs when the observed crude or unadjusted[a] association is exaggerated, or biased away from the null. Negative confounding occurs when the observed association is masked, or biased toward the null. There are multiple approaches to handle confounding, including randomization (e.g., randomizing participants in a clinical trial to a new treatment or a placebo), stratification (e.g., analyzing associations between the risk factor and the health outcome in key subgroups or strata of the potential confounder), matching (e.g., creating matched pairs that are similar on a number of potential confounding variables but differ in terms of the risk factor of interest), and multivariable statistical analysis. We review each approach separately and thereafter focus in detail on multivariable statistical analysis.

Effect modification, also called statistical interaction, occurs when the association between a risk factor and an outcome is different depending on groupings defined by a third variable (e.g., weaker in one group, stronger in one group, or in a different direction in one group compared to the other). Investigators often explore effect modification in key subgroups defined by a biological attribute hypothesized to be responsible for different relationships. Combining different estimates of association across subgroups when there is effect modification can result in an incorrect estimate of the true overall association in the population.

▶ 3.2 Approaches to Address Confounding

There are several different approaches to handle confounding, which can be broadly classified as approaches in the design or analysis phases of a study. In the design phase, investigators might use randomization, restriction, or matching to control confounding. In the analysis phase, investigators might

a A crude or unadjusted association is the association between the risk factor or exposure and the health outcome without consideration for any other variables.

use stratification or multivariable analysis. We describe each approach briefly and then discuss multivariable statistical analysis in more detail.

Randomization is a powerful technique to minimize confounding, where participants are randomly assigned to competing treatments or interventions. Randomization is accomplished by generating a randomization schedule[b] that assigns participants by chance to different treatments or interventions as they are enrolled in a study. Most randomized studies use blinding, whereby participants are not aware of the assigned treatment. This minimizes the placebo effect—a response that does not appear to be linked to an active treatment, but rather is attributable to a belief on the part of the participant that they were assigned the active treatment and therefore experience benefit. Randomization, in theory, balances the groups in terms of potential confounders, both known and unknown. Unfortunately, there are many situations where randomization is not possible. For example, suppose we wanted to evaluate the association between vaping and risk of heart attack in young adults. Ethically, we could not randomize participants to engage in vaping or not.

Restriction is a technique to control confounding whereby we more narrowly define the study sample in an attempt to minimize confounding. Continuing with the vaping example, we might be concerned that young adults who vape might also engage in other risky behaviors such as alcohol consumption. Thus, to control for confounding due to alcohol consumption, we could restrict the study sample to those who do not consume alcohol. This not only would remove alcohol consumption as a potential confounder but would also limit generalizability.[c]

Matching is a technique where exposed participants (e.g., those who vape) are matched to unexposed participants (e.g., those who do not vape) on a number of other characteristics that might be confounders (e.g., age, gender, alcohol consumption). Matching can be an effective way to control for confounding, but one of the challenges with matching is finding appropriate matches for each participant, and the more confounding variables there are, the more challenging this becomes.

There are additional techniques that can be applied during the statistical analysis phase to control for confounding. These include, but are not limited to, stratification and multivariable analysis. Stratification involves performing analyses separately in key subgroups defined by the confounder to evaluate the association between the exposure and health outcome. Continuing with the vaping example, we might evaluate the association between vaping and risk of heart attack separately in those who consume alcohol and those who do not. If the observed association in each subgroup, or strata, is similar

---

b A randomization schedule is created from a table of random numbers, available in most statistical textbooks, or generated by a statistical computing package.

c Generalizability refers to the extent to which results from the study sample extend to the population.

to that observed in the full sample (i.e., the crude or unadjusted association based on analysis with those consuming and not consuming alcohol combined), we would conclude that there is no impact of confounding by alcohol use. However, if the crude or unadjusted association between vaping and risk of heart attack is strong and the association is weak among those who consume alcohol and also among those who do not, we conclude that there is confounding by alcohol use. This is an example of positive confounding by alcohol use in that the crude analysis suggests a strong association between vaping and risk of heart attack, but when we minimize the impact of alcohol use by stratifying, we observe a weak association (i.e., alcohol use exaggerates the association between vaping and heart attack risk). Stratification, like restriction and matching, can be a useful technique to control for confounding, but it too is difficult when there are many potential confounders. More details on other approaches to handling confounding can be found in Aschengrau and Seage (2014).[5]

Multivariable statistical analysis is a widely applied analytical technique to control for confounding. There are many statistical approaches that are considered multivariable analyses; we focus on a few popular methods including multivariable linear regression, multivariable logistic regression, and survival analysis. In each analysis, we build statistical models (mathematical equations) relating the outcome to the key risk factor, or exposure of interest, and the potential confounding variables. **EXAMPLE 3-1** outlines the approach.

EXAMPLE 3-1

Investigators want to evaluate whether there is a statistically significant difference in knowledge of disease transmission risk among healthcare professionals completing a training course offered in-person compared to those completing an online training course. Participants select the course format that best fits their needs, and following completion of the course, each participant takes a test that is scored on a 100-point scale. A total of $n = 40$ participants selected the online option and their mean test score is 66.8 with a standard deviation of 13.1 points. A total of $n = 50$ participants selected the in-person option and their mean test score is 80.1 with a standard deviation of 11.5 points.

The difference in mean test scores is $(80.1 - 66.8) = 13.3$ points and a two independent samples t-test finds a statistically significant difference in mean test scores ($p < 0.001$), with those completing the in-person training course scoring statistically significantly higher than those completing the online training course.

Investigators look more closely at the participants who selected each course option and find that those who selected the online option are much younger by comparison to the in-person training group. The mean age of those who completed the online option is 20 years, whereas the mean age of those who completed the in-person option is 25 years. Therefore, the investigators are concerned about confounding by age. Age is related to the risk factor (i.e., there is a difference in age according to course format), and age could affect knowledge scores (e.g., older participants might have more life or work experience, which translates to more knowledge of disease transmission risk). A multivariable analysis is used to control for confounding by age. Specifically, multivariable modelling is used to predict knowledge scores, assuming the two groups have participants of similar age. **FIGURE 3-1** shows the crude and adjusted estimates of effect. Specifically, the crude, or unadjusted, difference in mean knowledge scores is 13.3 points (80.1 − 66.8 = 13.3). The adjusted difference in mean knowledge scores is 4.7 points (75.2 − 70.5 = 4.7), which is estimated at the mean age of the sample (22.8 years).

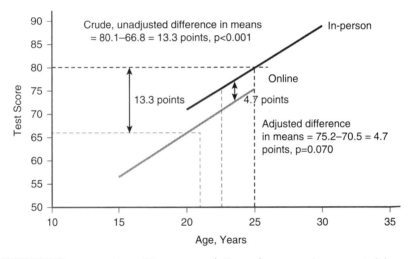

FIGURE 3-1 Test scores in participants completing online versus in-person trainings.

In Example 3-1, the crude or unadjusted comparison of mean test scores is highly statistically significant at $p < 0.001$. However, there is a difference in the ages of participants who selected each training option, which appears to exaggerate the crude association. If we adjust for age (i.e., we estimate the test scores in each group at age 22.8 years, the mean age in the sample), the difference in means is no longer statistically significant as $p = 0.070$ (assuming a 5% level of significance).

Multivariable analysis is a powerful technique that allows for adjustment of multiple confounding variables simultaneously. However, there are assumptions that must be carefully evaluated when using multivariable analysis. We discuss assumptions associated with each approach as we delve into the details of multivariable linear regression, multivariable logistic regression, and survival analysis. We first discuss data considerations for multivariable modelling.

▶ 3.3 Data Considerations for Multivariable Modelling

In Unit 1, we noted that the first step in any analysis is defining the outcome of interest. Examples of health outcomes include blood sugar level, body mass index (BMI), viral load, prevalent anxiety or substance use disorder, risk of developing cardiovascular disease, diabetes, or dementia. In every analysis, it is critical to correctly classify the outcome variable type as the appropriate statistical analysis depends on variable type. For multivariable modelling purposes, we focus on continuous (e.g., blood sugar level, BMI, viral load), dichotomous (e.g., prevalent anxiety, cardiovascular disease risk), and time-to-event outcomes (e.g., time to develop disease or time to death).

When the outcome is continuous, we generally focus on means (unless the outcome is highly skewed, in which case we focus on medians) and analyze multiple predictors of a continuous outcome using multivariable linear regression analysis. When the outcome is dichotomous, we focus on proportions and analyze multiple predictors of a dichotomous outcome using multivariable logistic regression analysis. When the outcome is time to developing disease or another health outcome, we use survival analysis methods. We will describe and illustrate multivariable linear regression, multivariable logistic regression and survival analysis methods in detail, but first discuss how to set up risk factors for these analyses.

Regression analysis techniques allow us to quantify the associations between multiple risk factors or predictors, considered simultaneously, and the outcome of interest. The risk factors or predictors can be continuous variables (e.g., age, heart rate) or dichotomous variables (e.g., family history of cardiovascular disease, treatment for hypertension). For analysis purposes, dichotomous variables are often coded as 0/1 with 1 assigned to those with the risk factor and 0 to those without. There are also instances when we wish to consider categorical or ordinal predictors. For example, self-reported physical health status is an example of an ordinal predictor variable where response options might be excellent, very good, good, fair, and poor. Ordinal variables can be numerically coded (e.g., 5 = excellent, 4 = very good, 3 = good, 2 = fair, 1 = poor) and analyzed as if they were continuous variables. This assumes, however, that one-unit differences (e.g., between very good and excellent or between poor and fair) are comparable across the range of the variable. Ordinal variables can also be modelled as distinct categories.

Categorical variables (e.g., race/ethnicity) must be handled very carefully in regression analysis. In order to include a categorical predictor (or an ordinal predictor as specific categories) in a regression analysis, we create dummy variables. Dummy variables are dichotomous variables that distinguish the different response categories. For example, suppose that participants in a study indicate which of the following best describes their race/ethnicity: black, Hispanic, white. (Note that there are many more racial/ethnic categories; we use an overly simplistic scheme here to briefly illustrate the technique for creating dummy variables.) If a categorical variable has k different response categories, then $(k - 1)$ dummy variables are required to distinguish the categories. In order to set up dummy variables, the investigator first needs to determine a reference category. This is the category against which the others will be compared. If interest lies in comparing whites to blacks and Hispanics to blacks in terms of the outcome of interest, then "black" is the reference category for race. If interest lies in comparing blacks to whites and Hispanics to whites in terms of the outcome of interest, then "white" is the reference category for race. Once the reference category is selected, the $(k - 1)$ dummy variables are then created to represent the other categories. Each of the dummy variables is coded 1 for participants classified in that category and 0 otherwise. **TABLE 3-1** illustrates this approach when (center columns) black is the reference category for race and when (right-hand columns) white is the reference category for race. The investigators need only choose one reference group; we show the two possibilities to illustrate the approach of creating dummy variables for different reference categories. If the reference is black race, then the dummy variables reflecting Hispanic and white race are entered into the multivariable analysis to model race/ethnicity. If the reference is white race, then the dummy variables reflecting black and Hispanic race are entered into the multivariable analysis to model race/ethnicity.

TABLE 3-1 Creating Dummy Variables for a Categorical Predictor

Recorded Race/ Ethnicity	Reference = Black Race		Reference = White Race	
	Hispanic	**White**	**Black**	**Hispanic**
Black	0	0	1	0
Hispanic	1	0	0	1
White	0	1	0	0

In all statistical analyses, it is important to assess missing data. In a particular application, we might be concerned about confounding due to age, gender, educational attainment, and household income. The latter, in particular, might be subject to extensive missing data as participants might not feel comfortable reporting their household income in a study. In most multivariable analyses, statistical computing packages use an approach called listwise deletion, which means that if a participant is missing any of the variables (the outcome, the primary risk factor or exposure, or any of the potential confounders), then that participant is excluded from the analysis.

If participants choose not to report household income and we include household income as a potential confounder, then anyone with missing income data is excluded from the analysis, which, in turn, reduces statistical power and precision and may also introduce bias. Thus, it is always important to check the completeness of all of the variables considered in a multivariable analysis.

Furthermore, in some analyses, a particular variable might be a predictor, and in another analysis, that same variable might be an outcome. For example, BMI might be the outcome in an analysis investigating predictors of obesity, and a predictor in an analysis of risk factors for incident cardiovascular disease. Investigators must always be clear in identifying variables for analysis and in specifying specific roles that variables are assumed to play.

▶ 3.4 Multivariable Linear Regression Analysis

Multivariable linear regression analysis is used to relate a single continuous outcome to one or more predictor variables. The predictors can be continuous, dichotomous, or a set of dummy variables. The assumptions for appropriate use of linear regression analysis include normality, similar variability in the outcome across the range of each predictor, a linear relationship between each predictor and the outcome, independence of observations (i.e., observations should be unrelated and not repeated measurements on the same individuals), and inclusion of a set of predictors that are not highly correlated with one another. There are scenarios where nonlinear associations are assumed between predictors and the outcome, or when some of these assumptions do not hold. Regression diagnostics procedures allow investigators to evaluate whether assumptions are likely met in a given application. For more details, see Armitage, Berry, and Matthews.[6]

Crude, Unadjusted Linear Regression Models

A crude or unadjusted linear regression model relates a single predictor, risk factor, or exposure to a continuous outcome. The crude or unadjusted linear regression model is: $\hat{y} = b_0 + b_1 x$, where $\hat{y}$ is the predicted or expected value of the outcome, x is the predictor (risk factor or exposure), b_0 is the y-intercept or the expected value of the outcome when the predictor is 0, and b_1 is the regression coefficient or the slope relating the predictor to the outcome. The regression coefficient, b_1, is almost always of the greatest interest as it quantifies the association between the predictor and the outcome. Specifically, b_1 is the expected change in the outcome, y, relative to a 1-unit change in the predictor, x. The actual value of the y-intercept, b_0, in and of itself, is not generally of interest, but is important for prediction (i.e., if we wish to estimate the outcome for a participant with a specific risk factor, x).

The regression coefficients (b_0 and b_1) are estimated using a technique called ordinary least squares, which is a technique that produces estimates for b_0 and b_1 for the equation of the line that best fits the data. The best-fitting line based on ordinary least squares is the line that minimizes the sum of the squared residuals. Residuals are differences between the observed outcomes, y, and the predicted outcomes, $\hat{y}$, based on the regression equation. Each participant has a residual, defined as $y - \hat{y}$, and the sum of the residuals in the study sample is 0. The estimates of the y-intercept and slope based on ordinary least squares produce a line that minimizes the sum of the squared residuals in the study sample. These estimates are usually produced by a statistical computing package, although there are formulas that can also be used to generate the estimates.

There are several ways to evaluate how well a statistical model fits the sample data. For linear regression, R^2 (R squared) is a popular measure of model fit. R^2 is the coefficient of determination and quantifies the percentage of variability in the outcome that is explained by the predictor(s). R^2 ranges from 0% to 100% with higher values indicating a better model fit. With most health-related outcomes, there are many factors that account for variability, and a low R^2 (e.g., 10%) may still be meaningful. Many statistical computing packages provide confidence intervals (CIs) around R^2 and also provide statistical tests for overall model fit to allow the investigator to test whether one or more predictors are statistically significantly associated with the outcome.

EXAMPLE 3-2 is an example of a linear regression analysis with BMI as the outcome of interest.

EXAMPLE 3-2

Suppose we wish to understand predictors of BMI in a large cross-sectional study of $n = 5,408$ adults between the ages of 50 and 75 years. We first examine the association between age, in years, and BMI, in kg/m². Note that both the outcome, BMI, and the predictor, age, are continuous. We use a statistical computing package to estimate the regression equation and the results are summarized in **TABLE 3-2**.

TABLE 3-2 Linear Regression Analysis Relating Body Mass Index (BMI) to Age

Predictor	Regression Coefficient	Standard Error	t	p	95% Confidence Interval
Intercept	26.648	0.530	50.28	<0.0001	(25.609, 27.687)
Age	−0.004	0.009	−0.47	0.6376	(−0.021, 0.013)

The estimated linear regression equation or model is: $\widehat{BMI} = 26.648 - 0.004$ age. The slope is estimated as $b_1 = -0.004$, which means that each additional year of age is associated with lower BMI by 0.004 kg/m^2. We can use the regression equation for prediction. For example, the predicted or expected BMI for an adult who is 50 years of age is $26.648 - 0.004(50) = 26.45$ kg/m^2. The predicted or expected BMI for an adult who is 65 years of age is $26.648 - 0.004(65) = 26.39$ kg/m^2. Regression equations can be used to predict health outcomes for participants within the range of the predictor, risk factor, or exposure studied. We should not, for example, use this regression equation to predict the BMI for an adult who is 30 or 80 years of age as we do not have data to support that the same association holds outside the age range of 50–75 years.

Based on the estimated slope ($b_1 = -0.004$), there is a weak, negative, or inverse association between age and BMI. The estimate of the slope has a standard error of 0.009, which allows us to generate a CI estimate for the slope using: point estimate ± margin of error, where the margin of error is the product of the t value for 95% confidence (with degrees of freedom, df $= n - 2$) and the standard error. The CI estimate can be represented as: point estimate $\pm t_{df} \times$ standard error. In this example, the 95% CI estimate for the slope is $-0.004 \pm 1.96 (0.009)$, -0.004 ± 0.017, or (-0.021 to 0.013). Because the CI includes the null value of 0, we conclude that the association between age and BMI is not statistically significant. Statistical computing packages also provide statistical tests for regression coefficients and in this example $p = 0.638$, which indicates that the population slope is not statistically significantly different from 0 and is not statistically significant (assuming a 5% level of significance) and, therefore, is consistent with our conclusion based on the CI estimate. The statistical computing package also produces an R^2 value and for this example $R^2 < 1\%$, indicating that less than 1% of the variability in BMI is explained by age.

Suppose that in the same study we also wish to understand the association between gender and BMI. We conduct a two independent samples t-test that finds the mean BMI among women to be 25.931 kg/m^2 and the BMI among men to be 26.946 kg/m^2, for a difference of 1.015 kg/m^2, which is statistically significant at $p < 0.0001$. We can also evaluate the association using linear regression analysis. First, we create an indicator of male gender (coded 1 for males and 0 for females), and then relate the indicator of male gender to BMI. (Note that we could also create an indicator of female gender coded 1 for females and 0 for males.) We use a statistical computing package to estimate the regression equation and the result is: $\widehat{BMI} = 25.931 + 1.015$ (male gender). The slope is estimated as $b_1 = 1.015$, which means that males, on average, have a higher BMI by 1.015 kg/m^2, compared to females. Note that this is precisely the difference we observed in the two independent samples t-test. Again, once we estimate a regression equation, it can be used for prediction. For example, the predicted or expected BMI for a female is $25.931 + 1.015 (0) = 25.931$ kg/m^2 and the predicted or expected BMI for a male is $25.931 + 1.015 (1) = 26.946$ kg/m^2. The statistical computing package also generates a 95% CI for the regression slope of (0.788,

1.243). Because the CI does not include the null value of 0, we conclude that the association between gender and BMI is statistically significant, which is consistent with our conclusion based on the p-value ($p < 0.0001$). There are many factors that are associated with BMI and most other health outcomes. After the next example, we explore how to estimate the association between BMI and multiple predictors or risk factors considered simultaneously.

EXAMPLE 3-3 is an example of a linear regression analysis with birthweight as the outcome of interest.

EXAMPLE 3-3

A study is conducted to evaluate risk factors for adverse pregnancy outcomes among $n = 486$ women with a history of premature birth. Women in the study self-identify as black, Hispanic, or white. We are interested in whether there are statistically significant differences in birthweights, in grams, among the three race/ethnicity groups. Because the predictor is a categorical variable, we create two dummy variables to distinguish the three race/ethnicity groups. We use white race as the referent, create indicators of black and Hispanic race, and use a statistical computing package to estimate the regression equation. The results are summarized in **TABLE 3-3**.

TABLE 3-3 Linear Regression Analysis Relating Birthweight to Mothers' Race/Ethnicity

Predictor	Regression Coefficient	Standard Error	t	p	95% Confidence Interval
Intercept	2,096.105	37.734	55.54	<0.0001	(2,021.944, 2,170.266)
Black race	335.672	55.377	6.06	<0.0001	(226.863, 444.481)
Hispanic race	−232.088	52.167	−4.45	<0.0001	(−334.590, −129.586)

The estimated regression equation or model is: $\overline{\text{birthweight}} = 2{,}096.105 + 335.672\,(\text{black race}) - 232.088\,(\text{Hispanic race})$. The regression slope associated with black race is estimated as 335.672, meaning that, on average, babies born to black mothers are 335.672 g heavier than babies born to white mothers (the referent). The regression slope associated with Hispanic race is estimated as −232.088, meaning that, on average, babies born to Hispanic mothers are 232.088 g lighter than babies born to white mothers. The 95% CI for the regression coefficient quantifying the difference in birthweight between babies born to black compared to white mothers is (226.863, 444.481), and the 95% CI for the regression coefficient quantifying the difference in birthweight between babies born to Hispanic compared to white mothers is (−334.590, −129.586). Neither CI includes the null value of 0, thus the differences in birthweights between babies born to black versus white mothers and Hispanic versus white mothers are statistically significant, which is consistent with the p-values ($p < 0.0001$ for both). The statistical computing package also produces $R^2 = 7.5\%$, meaning that 7.5% of the variability in birthweight is explained by race/ethnicity.

We can again use the regression equation for prediction. For example, the predicted or expected birthweight for a baby born to a white mother is $2{,}096.105 + 335.672\,(0) - 232.088\,(0) = 2{,}096.105$ g. In this example, we find statistically significant differences in birthweight by the mothers' race/ethnicity. It is important to always consider what else might explain any observed association, or lack thereof, between predictors and outcomes. There are other factors associated with birthweight that should be considered before attributing differences to race/ethnicity alone. Most important in any analysis that considers race/ethnicity as a predictor of any health outcome is the recognition of all that goes along with race—including racism and its many implications on health.

Multivariable, Adjusted Linear Regression Models

A multivariable linear regression model relates a set of predictors, risk factors, or exposures to a continuous outcome. The multivariable linear regression model is: $\hat{y} = b_0 + b_1 x_1 + b_2 x_2 + \ldots + b_p x_p$, where $\hat{y}$ is the predicted or expected value of the outcome, $x_1, x_2, \ldots, x_p$ is a set of p predictor, risk factor, or exposure variables that are continuous, dichotomous, or sets of dummy variables, and b_0 is the expected value of the outcome, y, when all of the predictors are 0. Each of the regression coefficients $b_1, b_2, \ldots, b_p$ quantifies associations between predictors and the outcome, adjusting for—or taking into account—other variables in the model. Specifically, b_1 is the expected change in the outcome, y, relative to a 1-unit change in the predictor, x_1, adjusting for other variables in the model.

Statistical computing software is used to estimate the regression coefficients in a multivariable model. Most packages also produce a number of other statistics including the standard errors of each regression coefficient, CI estimates for each regression coefficient, and t statistics to test whether regression coefficients are statistically significantly different from 0 (the null value). Additionally, most packages produce R^2, the coefficient of determination, which quantifies the percentage of variability in the outcome explained by the set of predictors and an overall analysis of variance (ANOVA) test that determines whether the set of predictors (the model) is statistically significant.

In interpreting a multivariable linear regression analysis, we first focus on the overall test to evaluate whether the set of predictors is statistically significant. If the overall test is statistically significant, we then move on to evaluate individual predictors. Individual predictors are tested for statistical significance after controlling for, or in the presence of, other variables using individual t-tests or CI estimates for each regression coefficient. In some scenarios, it is of interest to evaluate the relative importance of predictors and this is done by comparing p-values, with the smallest p-value being indicative of the most highly statistically significant predictor, or the absolute values of the test statistics associated with each regression coefficient, with the largest, in absolute value, being the most highly statistically significant. **EXAMPLE 3-4** illustrates this approach.

EXAMPLE 3-4

Returning to Example 3-2, we now estimate the association between BMI and age, gender, current smoking status, and systolic and diastolic blood pressures (SBP and DBP), considered simultaneously. The model is estimated using a statistical computing package and the results are summarized in **TABLE 3-4**.

TABLE 3-4 Multivariable Linear Regression Model Relating Body Mass Index (BMI) to Age, Gender, Current Smoking Status, and Systolic and Diastolic Blood Pressures

Predictor	Regression Coefficient	Standard Error	t	p	95% Confidence Interval
Intercept	20.001	0.726	27.57	<0.0001	(18.579, 21.423)
Age, years	−0.017	0.009	−1.79	0.0733	(−0.035, 0.002)
Male gender	0.930	0.114	8.17	<0.0001	(0.707, 1.153)
Current smoker	−1.209	0.120	−10.05	<0.0001	(−1.445, −0.973)
Systolic blood pressure	0.019	0.004	4.65	<0.0001	(0.011, 0.027)
Diastolic blood pressure	0.059	0.008	7.78	<0.0001	(0.044, 0.073)
$R^2 = 0.078$					

The statistical computing package produces an overall ANOVA test for the significance of the model with $F = 90.93$, $p < 0.0001$ indicating that the set of predictors is statistically significant. The overall R^2 is 7.8% indicating that 7.8% of the variability in BMI is explained by age, gender, current smoking status, SBP, and DBP.

In terms of individual predictors, all are highly statistically significant ($p < 0.0001$) with the exception of age, which does not reach statistical significance ($p = 0.0733$) assuming a 5% level of significance. Of the predictors considered, current smoking status is the most highly statistically significant (with $t = -10.05$), followed by gender ($t = 8.17$), diastolic blood pressure ($t = 7.78$), and SBP ($t = 4.65$). In some applications, it is possible to judge the relative importance of predictors based on the p-values. Here, the predictors that are statistically significant all have $p < 0.0001$, thus we use the t statistics to judge relative importance.

The interpretation of the regression coefficients is as follows. Current smokers, on average, have significantly lower BMI by 1.209 kg/m², compared to nonsmokers, adjusting for age, gender, SBP and DBP. Males, on average, have statistically significantly higher BMI, by 0.930 kg/m², compared to females, adjusting for age, current smoking status, SBP and DBP. Each additional increase of 1 mmHg of DBP is associated with a 0.059 kg/m² increase in BMI, adjusting for age, gender, current smoking status, and SBP, and each additional increase of 1 mmHg of SBP is associated with a 0.019 kg/m² increase in BMI, adjusting for age, gender, current smoking status, and DBP. With continuous predictors such as SBP and DBP, it is often more relevant to consider the impact of changes of 10 mmHg or another more clinically or practically relevant increment (e.g., a standard deviation increment) on the outcome. For example, each additional increase of 10 mmHg of DBP is associated with a 0.59 kg/m² increase in BMI, adjusting for age, gender, current smoking status, and SBP, and each additional increase of 10 mmHg of SBP is associated with a 0.19 kg/m² increase in BMI, adjusting for age, gender, current smoking status, and DBP.

In **EXAMPLE 3-5**, we use multivariable linear regression analysis to adjust for gestational age as a potential confounding variable in our analysis of the association between race/ethnicity and birthweight.

EXAMPLE 3-5

Returning to Example 3-3, we now estimate the association between birthweight, in grams, and the mothers' race/ethnicity adjusting for gestational age at birth. In this study sample of women with a history of premature birth, gestational age at birth is statistically significantly different by the mothers' race/ethnicity ($p < 0.0001$ based on ANOVA) as follows. The mean (standard deviation) gestational age of babies born to white mothers is 32.9 weeks

(continues)

EXAMPLE 3-5   (*continued*)

(4.4 weeks), compared to 34.7 weeks (3.6 weeks) for babies born to black mothers and 32.6 weeks (3.6 weeks) for babies born to Hispanic mothers. In addition, gestational age at birth is also statistically significantly associated with birthweight ($p < 0.0001$).[d] Thus, gestational age at birth is a potential confounder. We estimate a multivariable linear regression model relating the mothers' race/ethnicity to birthweight, adjusting for gestational age at birth. Again, women in the study self-identify as black, Hispanic, or white, and we create two dummy variables to distinguish the 3 race/ethnicity groups considering white race as the referent. We use a statistical computing package to estimate the regression equation and the results are summarized in **TABLE 3-5**.

TABLE 3-5 Multivariable Linear Regression Analysis Relating Birthweight to Mothers' Race/Ethnicity and Gestational Age at Birth

Predictor	Regression Coefficient	Standard Error	t	p	95% Confidence Interval
Intercept	−4254.245	147.004	−28.88	<0.0001	(−4534.092, −3956.397)
Black race	91.385	25.634	3.57	0.0004	(41.017, 141.752)
Hispanic race	−75.750	23.833	−3.18	0.0016	(−122.579, −28.921)
Gestational age, weeks	189.867	4.372	43.43	<0.0001	(181.277, 198.457)
	$R^2 = 0.812$				

d   The association between birthweight and gestational age is evaluated using correlation analysis (see Example 2-24) where we observed a correlation of $r = 0.898$ with a 95% CI of (0.879, 0.914) and $p < 0.0001$.

The statistical computing package produces an overall ANOVA test for the significance of the model with $F = 692.42$, $p < 0.0001$ indicating that the set of predictors is statistically significant. The overall R^2 is 81.2%, indicating that 81.2% of the variability in birthweight is explained by the mothers' race and gestational age at birth. Note that the crude, or unadjusted, model relating birthweight to mothers' race/ethnicity had $R^2 = 7.5\%$. Gestational age at birth is highly related to birthweight, increasing the variation explained to 81.2%.

Both race/ethnicity and gestational age at birth are statistically significant predictors of birthweight with gestational age at birth being more highly statistically significant ($p < 0.0001$). The crude or unadjusted regression coefficients quantifying the expected differences in birthweight between babies born to black compared to white mothers and Hispanic compared to white mothers were 335.672 and −232.088, respectively (see Table 3-3). The adjusted regression coefficients quantifying the expected differences in birthweight between babies born to black compared to white mothers and Hispanic compared to white mothers are 91.385 and −75.750, respectively (Table 3-5). After adjustment for gestational age at birth, the effects of race/ethnicity on birthweight remain statistically significant but are much lower, suggesting that some of the effect that we initially attributed to race/ethnicity in the crude analysis was actually due to gestational age at birth. This is an example of positive confounding where the crude estimates were exaggerated due to confounding by gestational age.

The magnitude of confounding can be quantified by taking the difference between the crude and adjusted estimates and dividing by the adjusted estimate. For example, the magnitude of confounding in our estimate of the effect of black race, compared to white race, on birthweight is (335.672 − 91.385)/91.385 = 2.67, meaning that the crude estimate is 2.67 times higher than the adjusted estimate. The magnitude of confounding in our estimate of the effect of Hispanic race, compared to white race, on birthweight is [−232.088 − (−75.750)]/−75.750 = 2.06, meaning that the crude estimate is 2.06-times higher than the adjusted estimate. Some investigators use a 10% rule to judge meaningful confounding, suggesting evidence of confounding when the crude and adjusted estimates differ by more than 10%. The important issue is whether the adjusted estimate implies a different association from a clinical or practical point of view. In Example 3-3, we would not want to report the crude estimates as they overstate the associations between different race/ethnicities categories and birthweight.

▶ 3.5 Multivariable Logistic Regression Analysis

Multivariable logistic regression analysis is used to relate a single dichotomous outcome to one or more predictor variables. It is possible to estimate a logistic regression model for an ordinal outcome; here, we focus on dichotomous outcomes. The predictors can be continuous, dichotomous, or a set of dummy variables. The assumptions for appropriate use of logistic regression analysis include that the outcome is dichotomous, that there is a linear association between each predictor and the log odds of the outcome (more on this below), independence of observations (i.e., observations should be unrelated and not repeated measurements on the same individuals), and inclusion of a set of predictors that are not highly correlated with one another.

Before we describe crude, or unadjusted, logistic regression models and multivariable, or adjusted, logistic regression models, we first introduce some notation. Logistic regression is applied when the outcome, y, is dichotomous. Examples of dichotomous outcomes might include prevalent hypertension, incident cardiovascular disease, or all-cause mortality. For analysis purposes, the outcome is usually coded as 0/1, with 1 assigned to those who have—or develop—the outcome of interest and 0 to those who do not have—or develop—the outcome of interest. The probability of outcome is defined as $p = P(y = 1)$ and the predicted or expected probability is denoted as $\hat{p}$ (p-hat). The logistic regression model is represented as: $\hat{p} = \dfrac{e^{b_0 + b_1 x_1 + b_2 x_2 + \ldots + b_p x_p}}{1 + e^{b_0 + b_1 x_1 + b_2 x_2 + \ldots + b_p x_p}}$. The specific functional form of the logistic regression model ensures that predicted values are between 0 and 1, the defined range of a probability. The same model can be represented in another way: $\ln\left(\dfrac{\hat{p}}{1 - \hat{p}}\right) = b_0 + b_1 x_1 + b_2 x_2 + \ldots + b_p x_p$. In this more popular form, the predictor (right-hand) side of the equation looks like the more familiar multivariable linear regression model as a linear combination of the predictors. The outcome (left side) of the equation is the log odds of the outcome, also called the logit of the outcome. Each regression coefficient (i.e., b_1, b_2, ..., b_p) in a logistic regression model quantifies the expected change in the logit or log odds of the outcome relative to a 1-unit change in the predictor, taking into account other variables in the model. If we exponentiate a regression coefficient, we get an odds ratio (OR) which corresponds to a 1-unit change in that predictor. We illustrate the interpretation of regression coefficients in crude, or unadjusted, and multivariable, or adjusted, logistic regression models in the following sections.

Crude, Unadjusted Logistic Regression Models

The crude, or unadjusted, logistic regression model relates a single predictor, risk factor, or exposure to a dichotomous outcome. The crude, or unadjusted, logistic regression model is: $\ln\left(\dfrac{\hat{p}}{1 - \hat{p}}\right) = b_0 + b_1 x$, where $\ln\left(\dfrac{\hat{p}}{1 - \hat{p}}\right)$ is the predicted or expected log odds of the outcome, x is the predictor, risk factor, or exposure, b_0 is the predicted or expected log odds of the outcome when the predictor is 0, and b_1 is the regression coefficient relating the predictor to the log odds of the outcome. If we exponentiate b_1 [i.e., $\exp(b_1)$], we get the estimated OR corresponding to a 1-unit change in the predictor, x.

Similar to linear regression, the actual value of the y-intercept, b_0, is not generally of interest, but is important for prediction (i.e., if we wish to estimate the probability that an individual with a specific risk factor, x, will have, or develop, the outcome of interest).

The regression coefficients in a logistic regression model are estimated using an optimization technique, called maximum likelihood estimation, that is implemented with statistical software. This technique produces estimates for b_0 and b_1 that maximize the likelihood of observing the data that were realized. Most statistical computing packages produce estimates of each regression coefficient along with standard errors, CI estimates, and chi-square test statistics to evaluate whether associations are statistically significant (i.e., whether regression coefficients are statistically significantly different from 0, the null value). Statistical computing packages also produce estimates of ORs ($\widehat{OR}$) and CI estimates for the ORs that can be used to evaluate whether associations are statistically significant (i.e., if ORs are statistically significantly different from 1, the null value).

There are several ways to evaluate how well a logistic regression model fits the sample data. First, most statistical computing packages produce an overall statistical test of the model. This is a chi-square statistic (sometimes labelled the likelihood ratio chi-square) and determines whether one or more predictors is/are statistically significantly associated with the outcome. It is also possible to produce an R^2 for logistic regression, but it does not have the same interpretation as in linear regression and it also has an upper limit less than 1, so well-fitting models can result in low values of R^2, which makes it a difficult measure to interpret. More popular measures of model fit for logistic regression are discrimination and calibration, which are described in detail below.

Discrimination measures how well a statistical model distinguishes participants with, and without, the outcome of interest. Discrimination is summarized in a c statistic, which is equivalent to the area under the receiver operating characteristic (ROC) curve (see Figure 2-1). The c statistic is interpreted as the probability that the statistical model produces a higher predicted probability of outcome in participants who actually have the outcome compared to participants who do not have the outcome. The c statistic ranges from 0.5 to 1, with higher values indicative of better discrimination, and c statistics above 0.7 generally suggest that a model has good discrimination.[7]

Calibration measures how well-predicted probabilities of outcome status (i.e., $\hat{p}$, based on the statistical model) agree with actual outcomes. It is possible to evaluate whether a model is well calibrated using a statistical test. The test statistic is constructed by generating predicted probabilities of outcome for each participant in the study and organizing these predicted probabilities into deciles. In each decile, the predicted number of participants with the outcome is compared to the actual number of outcomes observed.

If the predicted and actual numbers are close in value, this suggests that the model is well calibrated. A chi-square statistic is produced and statistical significance suggests lack of fit (i.e., if $p < 0.05$, the model is not a good fit to the data). A popular version of this test is called the Hosmer and Lemeshow test.[7] It is also possible to use more informal assessments of calibration based on graphical analysis comparing predicted and expected outcomes in each decile.

There are other measures of model fit that are relevant depending on the specific purpose of the multivariable model, such as whether the model is designed for prediction purposes or to evaluate the prognostic value of a novel measure (i.e., does the addition of a new marker improve prediction of the outcome). More details can be found in Steyerberg et al.[8] Most statistical computing packages produce estimates of R^2 (which are not as useful for assessing fit of logistic regression models), c statistics to evaluate discrimination, and tests for model calibration as part of their standard output.

EXAMPLE 3-6 is an example of a logistic regression analysis with premature birth as the outcome of interest.

EXAMPLE 3-6

A study evaluates whether there is an association between self-reported alcohol consumption during pregnancy and premature birth (defined as the birth of a baby before the 37th week of pregnancy) in a sample of $n = 3,896$ pregnant women. The outcome of interest is premature birth, a dichotomous variable, and the predictor is an indicator of alcohol consumption during pregnancy (coded 1 for women who report consuming alcohol during pregnancy and 0 for those who report they do not consume alcohol during pregnancy). In the study sample, 321 (8.2%) women had their babies prematurely. We use a statistical computing package to estimate the logistic regression model and the results are summarized in **TABLE 3-6**.

TABLE 3-6 Logistic Regression Analysis Relating Premature Birth to Self-Reported Alcohol Consumption During Pregnancy

Predictor	Regression Coefficient	Standard Error	χ^2	p	95% Confidence Interval	Odds Ratio	95% Confidence Interval
Intercept	−2.425	0.059	1,696.50	<0.0001	(−2.542, −2.311)	—	
Alcohol during pregnancy	1.113	0.430	6.70	0.0096	(0.190, 1.902)	3.043	(1.210, 6.700)

The estimated regression equation is: $\ln\left(\dfrac{\widehat{\text{premature}}}{1-\widehat{\text{premature}}}\right) = -2.425 +$ 1.113 alcohol in pregnancy. The regression coefficient of interest is $b_1 = 1.113$, which indicates that women who report consuming alcohol during pregnancy have 1.113 times the log odds of premature birth compared to women who do not consume alcohol. If we exponentiate the regression coefficient we get the $\widehat{\text{OR}}$, $\exp(1.113) = 3.043$. Most statistical computing packages generate $\widehat{\text{OR}}$s and 95% CIs for the ORs as part of the output (see Table 3-6). Women who report consuming alcohol during pregnancy have 3.043 times the odds of premature birth compared to women who do not consume alcohol, and this is statistically significant as the 95% CI does not include the null value of 1, and $p = 0.0096$. This model is statistically significant with a model (likelihood ratio) $\chi^2 = 5.376$, $p = 0.0204$, but the discrimination is low, with $c = 0.507$. In the next section, we investigate whether additional predictors improve model discrimination.

EXAMPLE 3-7 is another example of a logistic regression analysis with obesity as the outcome of interest.

EXAMPLE 3-7

A study evaluates whether there is an association between regular physical activity, defined as 60 minutes or more at least 5 days per week, and obesity, defined as BMI $\geq$ 30 kg/m^2 in a sample of $n = 5{,}408$ adults between the ages of 50 and 75 years. The outcome of interest is an indicator of obesity, a dichotomous variable, and the predictor is an indicator of regular physical activity (coded 1 for adults who engage in 60 minutes or more at least 5 days per week and 0 for those who do not exercise regularly). We use a statistical computing package to estimate the logistic regression model and the results are summarized in **TABLE 3-7**.

TABLE 3-7 Logistic Regression Analysis Relating Obesity to Regular Physical Activity in Adults 50–75 Years of Age

Predictor	Regression Coefficient	Standard Error	χ^2	p	95% Confidence Interval	Odds Ratio	95% Confidence Interval
Intercept	−1.414	0.042	1,139.90	<0.0001	(−1.497, −1.332)	—	
Physically active	−0.490	0.082	35.80	<0.0001	(−0.652, −0.331)	0.613	(0.521, 0.718)

The estimated regression equation is: $\ln\left(\dfrac{\widehat{obese}}{1-\widehat{obese}}\right)=-1.414-0.490$ physically active. The regression coefficient of interest is $b_1=-0.490$, which indicates that adults who engage in regular physical activity have 0.490 times the log odds of obesity compared to adults who do not. If we exponentiate the regression coefficient we get the $\widehat{OR}$, $\exp(-0.490)=0.613$. Thus, adults who engage in regular physical activity have 0.613 times the odds of obesity compared to adults who do not exercise regularly. Equivalent to this is adults who engage in regular physical activity have 38.7% [100 × (1 − 0.613)] lower odds of obesity compared to adults who do not exercise regularly. The association between regular physical activity and obesity is statistically significant as the 95% CI does not include the null value of 1, and $p < 0.0001$. This model is statistically significant with a model $\chi^2 =$ 37.792, $p < 0.0001$, but the discrimination is again low, with $c = 0.551$. In the next section, we investigate whether additional predictors improve model discrimination.

Multivariable, Adjusted Logistic Regression Models

A multivariable logistic regression model relates a set of predictors, risk factors, or exposures to a dichotomous outcome. The multivariable logistic regression model is: $\ln\left(\dfrac{\hat{p}}{1-\hat{p}}\right)=b_0+b_1x_1+b_2x_2+\ldots+b_px_p$, where $\ln\left(\dfrac{\hat{p}}{1-\hat{p}}\right)$ is the predicted or expected log odds of the outcome, $x_1, x_2, \ldots, x_p$ is a set of p predictor, risk factor, or exposure variables that are continuous, dichotomous, or a set of dummy variables, and b_0 is the predicted log odds of the outcome when all of the predictors are 0. The regression coefficients, $b_1, b_2, \ldots, b_p$, quantify associations between the predictors and the outcome, adjusting for—or taking into account—other variables in the model. Specifically, b_1 is the expected change in the log odds of the outcome relative to a 1-unit change in the predictor, with x_1, adjusting for other variables in the model. If we exponentiate b_1 [i.e., $\exp(b_1)$], we get an $\widehat{OR}$ corresponding to a 1-unit change in the predictor, x_1, and adjusting for other variables in the model.

The regression coefficients in a multivariable logistic regression model are estimated using maximum likelihood estimation that is implemented with statistical software. The technique produces estimates for $b_0, b_1, b_2, \ldots,$ b_p that maximize the likelihood of observing the data that were realized. Most statistical computing packages produce estimates of each regression coefficient along with standard errors, CI estimates, and chi-square test statistics to evaluate whether associations are statistically significant (i.e., if the regression coefficient is statistically significantly different from 0, the null value), $\widehat{OR}$s and CIs for ORs to allow for assessment of statistical significance. A predictor is statistically significant if the 95% CI for the OR does not include 1, the null value, after considering the other variables in the model.

In interpreting a multivariable logistic regression model, we focus first on the overall test to evaluate whether the set of predictors is statistically significant. If the overall test is statistically significant, we then move on to evaluate individual predictors. Individual predictors are tested for statistical significance, after controlling for—or in the presence of—other variables, using individual chi-square tests for each regression coefficient, CI estimates for each regression coefficient, or CI estimates for ORs. In some scenarios, it is of interest to evaluate the relative importance of predictors and this is done by comparing *p*-values (with the smallest *p*-value indicative of the most highly statistically significant predictor) or the values of the test statistics associated with each regression coefficient (with the largest being the most highly statistically significant). **EXAMPLE 3-8** illustrates this approach.

EXAMPLE 3-8

Returning to Example 3-6, we now estimate the association between self-reported alcohol consumption during pregnancy and premature birth (defined as birth of the baby before the 37th week of pregnancy) in a sample of $n = 3,896$ pregnant women, adjusting for smoking during pregnancy, gestational diabetes, and mothers' age in years. We use a statistical computing package to estimate the multivariable logistic regression model and the results are summarized in **TABLE 3-8**.

TABLE 3-8 Multivariable Logistic Regression Analysis Relating Premature Birth to Self-Reported Alcohol Consumption and Smoking During Pregnancy, Gestational Diabetes Status, and Mothers' Age

Predictor	Regression Coefficient	Standard Error	χ^2	*p*	95% Confidence Interval	Odds Ratio	95% Confidence Interval
Intercept	−3.574	0.293	148.58	<0.0001	(−4.149, −3.000)	—	—
Alcohol during pregnancy	0.919	0.435	4.47	0.0345	(0.067, 1.771)	2.506	(1.069, 5.876)
Smoking during pregnancy	0.516	0.263	3.85	0.0499	(0.001, 1.032)	1.675	(1.000, 2.806)
Gestational diabetes	1.072	0.407	6.94	0.0084	(0.275, 1.870)	2.922	(1.316, 6.487)
Mothers' age, years	0.037	0.009	15.59	<0.0001	(0.019, 0.056)	1.038	(1.019, 1.058)

In Example 3-8, we use a multivariable logistic regression model to adjust for smoking during pregnancy, gestational diabetes, and mother's age in years as potential confounding variables in our analysis of the association between alcohol consumption during pregnancy and premature birth.

The statistical computing package produces an overall test for the model, model $\chi^2 = 30.998$, $p < 0.0001$, which indicates that the set of predictors is statistically significant. The primary risk factor, alcohol consumption during pregnancy, remains statistically significant after adjustment for other variables. All of the individual predictors are statistically significant at the 5% significance level. Mothers' age, in years, is the most highly statistically significant predictor ($p < 0.0001$), followed by gestational diabetes ($p = 0.0084$), alcohol use during pregnancy ($p = 0.0345$), and smoking during pregnancy ($p = 0.0499$).

The adjusted $\widehat{OR}$ associated with alcohol consumption during pregnancy is 2.506 with a 95% CI (1.069, 5.876). Thus, women who report consuming alcohol during pregnancy have 2.506 times the odds of premature birth compared to women who did not consume alcohol during pregnancy, adjusting for smoking during pregnancy, gestational diabetes, and mothers' age. The other predictors are interpreted as follows. Women who report smoking during pregnancy have 1.675 times the odds of premature birth compared to women who did not, adjusting for alcohol consumption during pregnancy, gestational diabetes, and mothers' age, and this association is statistically significant ($p = 0.0499$). Women with gestational diabetes have 2.922 times the odds of premature birth compared to women who do not, adjusting for alcohol consumption and smoking during pregnancy, and mothers' age, and this association is statistically significant ($p = 0.0084$). Older mothers have statistically significantly higher odds of premature birth ($p < 0.0001$), with each additional year of age associated with a 3.8% higher odds (or 1.038 times the odds) of premature birth, adjusting for alcohol consumption during pregnancy, smoking during pregnancy, and gestational diabetes.

We can estimate the magnitude of confounding in our estimate of the effect of alcohol consumption during pregnancy and premature birth by taking the difference between the crude and adjusted $\widehat{OR}$s and dividing the result by the adjusted $\widehat{OR}$. Specifically, the magnitude of confounding in our estimate of the effect of alcohol consumption during pregnancy on premature birth is $(3.043 - 2.506)/2.506 = 0.214$, meaning that the crude estimate is 21% higher than the adjusted estimate. Using the 10% rule, we conclude that there is evidence of positive confounding.

The discrimination of the model remains low, with $c = 0.570$. The test for calibration of this model is statistically significant with $\chi^2 = 1886.35$ and $p < 0.0001$, suggesting that the model is not a good fit to the data. There may be other factors, perhaps some not measured in this study, which are important predictors of premature birth.

In **EXAMPLE 3-9**, we use a multivariable logistic regression model to adjust for age, gender, SBP, DBP, and total serum cholesterol as potential confounding variables in our analysis of the association between regular physical activity and obesity.

EXAMPLE 3-9

Returning to Example 3-7, we now estimate the association between regular physical activity, defined as 60 minutes or more at least 5 days per week, and obesity, defined as BMI $\geq$ 30 kg/m^2 in a sample of $n = 5{,}408$ adults between the ages of 50 and 75 years, adjusting for age, gender, SBP, DBP and total serum cholesterol level. We use a statistical computing package to estimate the multivariable logistic regression model and the results are summarized in **TABLE 3-9**.

TABLE 3-9 Multivariable Logistic Regression Analysis Relating Obesity to Regular Physical Activity in Adults aged 50–75 Years, Adjusted for Age, Gender, Systolic and Diastolic Blood Pressure, and Total Serum Cholesterol Level

Predictor	Regression Coefficient	Standard Error	χ^2	p	95% Confidence Interval	Odds Ratio	95% Confidence Interval
Intercept	−4.500	0.505	79.51	<0.0001	(−5.494, −3.515)	—	—
Regular physical activity	−0.502	0.084	35.57	<0.0001	(−0.668, −0.338)	0.606	(0.513, 0.713)
Age, years	−0.011	0.001	3.17	0.0752	(−0.023, 0.001)	0.989	(0.977, 1.001)
Male gender	0.001	0.076	0.01	0.9848	(−0.148, 0.151)	1.001	(0.862, 1.163)
Systolic blood pressure	0.010	0.003	14.85	<0.0001	(0.005, 0.015)	1.010	(1.004, 1.015)
Diastolic blood pressure	0.028	0.005	32.65	<0.0001	(0.018, 0.037)	1.029	(1.019, 1.038)
Total cholesterol	0.0002	0.001	0.07	0.7983	−0.002, 0.002)	1.000	(0.998, 1.002)

The statistical computing package produces an overall test for the model, model $\chi^2 = 199.348$, $p < 0.0001$, which indicates that the set of predictors is statistically significant. The primary risk factor, regular physical activity, remains statistically significant after adjustment for other variables. SBP and DBP are highly statistically significant predictors of obesity, with $p < 0.0001$ for each. Age does not reach statistical significance at the 5% level, with $p = 0.0752$. Male gender and total serum cholesterol levels are not statistically significant in the presence of the other variables considered.

The adjusted $\widehat{OR}$ associated with regular physical activity is 0.606 with a 95% CI (0.513, 0.713). Thus, adults who engage in regular physical activity have 0.606 times the odds of obesity compared to adults who do not, adjusting for age, gender, SBP, DBP, and total serum cholesterol level. Equivalent to this is adults who engage in regular physical activity have 39.4% $[100 \times (1 - 0.606)]$ lower odds of obesity compared to adults who do not exercise regularly, adjusting for age, gender, SBP, DBP, and total serum cholesterol level. The other predictors are interpreted as follows. Each additional 1 mmHg increase in SBP is associated with 1% higher odds (or 1.010 times the odds) of obesity, adjusting for regular physical activity, age, gender, DBP, and total serum cholesterol level. Each additional 1 mmHg increase in DBP is associated with 2.9% higher odds (or 1.029 times the odds) of obesity, adjusting for regular physical activity, age, gender, SBP, and total serum cholesterol level. With continuous predictors, such as SBP and DBP, it is often more relevant to consider the impact of changes of 10 mmHg or another more clinically relevant increment (e.g., a standard deviation increment) on obesity. For example, each additional increase of 10 mmHg of SBP is associated with 11% higher odds (1.105 times the odds) of obesity, adjusting for regular physical activity, age, gender, DBP, and total serum cholesterol level.[e] Each additional 10 mmHg increase in DBP is associated with 33% higher odds (or 1.328 times the odds) of obesity, adjusting for regular physical activity, age, gender, SBP, and total serum cholesterol level.

Our primary predictor is regular physical activity and we estimate the magnitude of confounding by taking the difference between the crude and adjusted $\widehat{OR}$s and dividing the result by the adjusted $\widehat{OR}$. Specifically, the magnitude of confounding in our estimate of the effect of regular physical activity on obesity is $(0.613 - 0.606)/0.606 = 0.012$, meaning that the crude estimate is 1.2% higher than the adjusted estimate, which does not meet the 10% rule. Thus, there is minimal evidence of confounding due to age, gender, SBP, DBP, and total serum cholesterol level.

The discrimination of the model remains low, with $c = 0.640$. The test for calibration of this model is statistically significant, with $\chi^2 = 4,801.07$ and $p < 0.0001$, suggesting that the model is not a good fit to the data. Here, too, there may be other factors, perhaps some not measured in the study, which are important predictors of obesity.

---

e To estimate the OR relative to a c unit increment, we exponentiate ($c \times$ regression coefficient). For example, the estimate of the OR relative to a 10-unit increase in systolic blood pressure is $\exp(10 \times 0.010) = 1.105$.

▶ **3.6 Survival Analysis**

Survival analysis methods are statistical techniques applicable when the outcome is time to a particular event, such as the time it takes to develop a disease or time until death. Survival analysis techniques were initially developed to model mortality, hence the name, but have much wider applicability and are more generally called time-to-event analysis. The key feature in survival or time-to-event analysis is censoring. Specifically, participants are followed for a specified, clinically relevant, follow-up period (e.g., 10 years), and some develop the event of interest, some do not, others may drop out of the study, and some might be lost to follow-up (i.e., the investigators lose contact with the participant). For participants who experience the event of interest during the follow-up period, we measure their time to event. For those who do not, we measure follow-up time, which is the last known time when the participant was event-free. The latter is called a censored time. Censored times are called right-censored when the follow-up period ends or when a participant leaves the study before experiencing the event of interest. Censored times are called left-censored when the event of interest occurs before the study begins and interval-censored times are those where the event of interest occurs within a particular time window, but we do not know exactly when. The most common type of censoring is right-censoring. The outcome in survival analysis is based on two components: an indicator of event status (whether the event occurs within the follow-up period or not) and time (either time-to-event for those who experience the event or follow-up time, a censored time, for those who do not). Here, we describe three popular analyses of survival or time-to-event data: estimating a survival function, comparing survival functions in two or more independent groups, and multivariable modelling using Cox proportional hazards regression analysis to relate a time-to-event outcome to one or more predictors or risk factors.

Estimating a Survival Function

Data for survival or time-to-event analysis for each participant include when they enter or begin a study, when they suffer the event of interest, or if they do not, the time at which they are last known to be event-free. For each participant, we compute their time on study and their status at the end of this time (i.e., either an event occurred or they are censored). A basic survival analysis includes the creation of a life table that organizes follow-up time into discrete periods (e.g., for a 10-year follow-up, we might create 5 periods of 2 years each) and in each period, the total number of participants at risk, the number who suffer the event of interest, and the number who are censored are summarized. The probability of death (or the probability of developing the outcome of interest) is computed as is the survival probability computed by subtracting the probability of death (or the outcome of interest) from 1. The Kaplan-Meier method is a variation on the life table approach and is the most widely used approach to estimate a survival function. The Kaplan-Meier approach creates time periods that end at the actual times that events occur in the study sample rather than defining fixed time periods.[9] Within each period, survival probabilities are computed. The Kaplan-Meier approach addresses a particular issue with the life table approach in that survival probabilities can be different depending on how the time periods are organized. More details on the Kaplan-Meier approach can be found in Rich et al.[10]

A survival function is a graphical display of survival probabilities against time and is sometimes called a Kaplan-Meier curve. The survival function shows a survival probability of 1 at time 0 and steps downward. If the survival function drops sharply, outcome events tend to occur early and often in follow-up. If the survival function remains flat and drops slowly, few outcome events occur over time. If there are participants who survive past the end of the follow-up period, the survival function does not reach 0 at the end of the follow-up period. When the time-to-event outcome is time-to-incident disease (e.g., incident stroke or cancer) as opposed to time-to-death, investigators sometimes display the cumulative incidence of disease rather than survival probability. A cumulative incidence curve plots risk of disease against time. If the cumulative incidence curve rises quickly, this suggests that outcome events occur early and often, whereas if the cumulative incidence curve remains flat for an extended period, this suggests that outcome events occur infrequently and later in time. Cumulative incidence at each time point is computed by subtracting the Kaplan-Meier survival probability from 1, as long as there are no competing risks. A competing risk is another outcome that makes it impossible for the event of interest to occur. For example, we might be evaluating time to develop Alzheimer's disease in a study of older adults. Over the course of the follow-up period, some participants might die due to cardiovascular disease, which complicates the analysis of time to develop Alzheimer's disease. Special attention to these details is required for analysis of time-to-event data with competing risks. Details can be found in Austin et al.[11]

There are several statistics that are reported based on a survival function. The first is an estimate of median survival time in a study sample (i.e., the time at which 50% of the sample is free of the event of interest), which is a very commonly reported statistic. It can also be of interest to estimate survival probabilities at key time points, for example, 1 year, 5 years, or 10 years. **EXAMPLE 3-10** illustrates this approach.

EXAMPLE 3-10

A sample of $n = 35$ patients with metastatic cervical cancer are treated with a specific chemotherapy protocol at a local hospital. Each receives the chemotherapy and is followed for up to 4 years. Many of the participants die over the course of the follow-up period and their time of death is recorded. **FIGURE 3-2** is the survival function estimated using the Kaplan-Meier method based on the experiences of the $n = 35$ participants in the study.

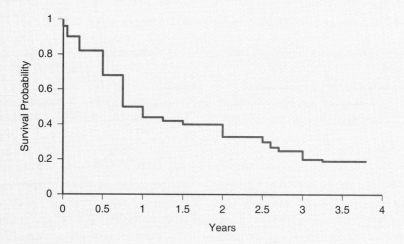

FIGURE 3-2 Survival over 4-year follow-up period in patients with metastatic cervical cancer.

Sometimes survival functions are shown as step functions, as in Figure 3-2, where each step represents times at which patients in the sample experience the event of interest, and sometimes survival functions are presented as smoothed curves. In either case, time is shown on the x-axis and starts at time 0, which usually represents the start of the study or, in this case, the start of treatment. Survival probability is shown on the y-axis and is interpreted as the probability of surviving a particular time period. For example, the probability of surviving 6 months (0.5 years) is 70%, the probability of surviving 1 year is 45%, and the probability of surviving 3 years is 25%.

Some investigators show numbers just below the x-axis to convey the numbers of patients at risk over time. This starts with the total sample size at time 0 and decreases over time as participants suffer events, drop out of the study, or are censored. Still others show CI estimates or confidence bands around the survival function to convey that these are estimates and that there is a range of plausible survival probabilities at each time point.

Comparing Survival in Two or More Independent Groups

As we discussed in Unit 2, it is often of interest to compare two or more independent groups in terms of a continuous outcome using a two independent samples *t*-test when there are exactly two groups, or ANOVA when there are more than two groups, or to compare categorical, ordinal, or dichotomous outcomes in two or more independent groups using chi-square tests. In this section, we illustrate how to compare time-to-event data in two or more independent groups. The groups might be defined by the investigator (e.g., participants randomized to one treatment or another in a clinical trial) or based on a particular condition or attribute (e.g., patients under 30 years of age, patients between 30 and 45 years of age, and patients over 45 years of age). The outcome of interest in each group is time-to-event and we wish to compare survival functions among groups.

A popular test of hypotheses to compare two or more independent survival functions is the logrank test. The null hypothesis in the logrank test reflects the situation where the survival functions are equal across comparison groups. The research hypothesis reflects the situation where the survival functions are different across comparison groups. There are several variations of the logrank test statistic. Most statistical computing packages offer several options that differ in terms of how they address early events versus later events such that they weight more heavily events that occur earlier as opposed to later, but with larger samples the conclusions are usually consistent regardless of the specific test applied. The logrank chi-square statistic compares observed numbers of events to what would have been expected if there were no differences among groups in terms of survival, and statistical significance is summarized in a *p*-value, the probability of observing a chi-square statistic as, or more extreme than, that observed. **EXAMPLE 3-11** illustrates the interpretation of a logrank test. More details on the computation of the logrank statistic can be found in Bland and Altman (2004).[12]

EXAMPLE 3-11

Continuing with Example 3-10, suppose we have a second sample of $n = 22$ patients with metastatic cervical cancer who are treated with radiation therapy, and we wish to compare survival outcomes in those treated with radiation therapy compared to those treated with chemotherapy. **FIGURE 3-3** shows the estimated survival functions for each treatment group using the Kaplan-Meier method.

(continues)

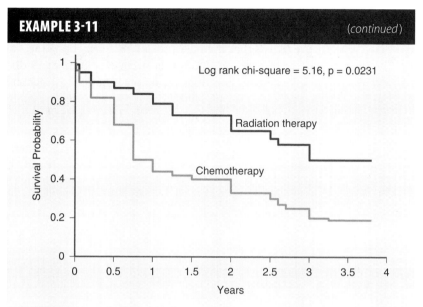

FIGURE 3-3 Survival over 4-year follow-up in patients with metastatic cervical cancer treated with chemotherapy as compared to radiation therapy.

Figure 3-3 shows that survival is worse for patients who are receiving chemotherapy compared to those receiving radiation therapy. For example, for those receiving chemotherapy, the probability of surviving 6 months (0.5 years) is 70% compared to 90% for those receiving radiation therapy. For those receiving chemotherapy, the probability of surviving 1 year is 45% compared to 80% for those receiving radiation therapy, and for those receiving chemotherapy, the probability of surviving 3 years is 25% compared to 53% for those receiving radiation therapy.

We conduct a logrank test using a statistical computing package and observe a logrank χ^2 statistic = 5.16, with $p = 0.0231$. There is a statistically significant difference in survival for patients with metastatic cervical cancer receiving chemotherapy compared to those receiving radiation therapy, with those receiving radiation therapy experiencing better survival. As we have discussed with previous examples in this unit, the question is whether there are other explanations for this observed difference. Are the patients who received chemotherapy comparable to the patients who received radiation therapy? Are there other risk factors that might be confounding the observed association between treatment and survival probability? Multivariable methods allow us to address these important questions.

Cox Proportional Hazards Regression Analysis

Cox proportional hazards regression analysis is a popular method used to relate a time-to-event outcome to one or more predictor variables.

The predictors can be continuous, dichotomous, or a set of dummy variables. Time-to-event analysis can also accommodate time-varying predictors. These are predictors that change over the course of the follow-up period, such as blood pressure, onset of other diseases, exposure to different environmental conditions, initiation or termination of medical treatments, and so on. Time-varying predictors can be accommodated in Cox proportional hazards regression models. The term "hazard" refers to the instantaneous risk that a participant experiences the outcome of interest, given they have not yet experienced it. A hazard ratio is computed by dividing the hazard in the intervention, treatment, or exposed group by the hazard in the control, standard care, or unexposed group. In practice, many people interpret hazard ratios as risk ratios, although they are technically not the same. Risk ratios summarize risks that outcomes occur by the end of the study. In contrast, hazard ratios are specific to a particular time.

Cox proportional hazards regression is a semiparametric method that makes more assumptions than a nonparametric method[13] but fewer than a parametric method. Specifically, the assumptions for appropriate use of Cox proportional hazards regression include that hazard ratios are constant over time and that there is a linear association between each predictor and the log of the hazard, independence of observations (i.e., observations should be unrelated and not repeated measurements on the same individuals), and inclusion of a set of predictors that are not highly correlated with one another.

The Cox proportional hazards regression model is: $\log h(t) = \log h_0(t) + b_1 x_1 + b_2 x_2 + \ldots + b_p x_p$, where $h(t)$ is the hazard function and $h_0(t)$ is the baseline hazard (both are functions of time), $b_1, b_2, \ldots, b_p$ are regression coefficients associated with $x_1, x_2, \ldots, x_p$, the set of predictors or risk factors. Each regression coefficient (i.e., $b_1, b_2, \ldots, b_p$) quantifies the expected change in the log of the hazard at time t relative to a 1-unit change in the predictor and is estimated using maximum likelihood estimation. We exponentiate regression coefficients to produce estimates of hazard ratios that correspond to 1-unit changes in each predictor. We illustrate the interpretation of regression coefficients in crude, or unadjusted, and multivariable, or adjusted, Cox proportional hazards regression models in the following examples.

There are several ways to evaluate how well a Cox proportional hazards regression model fits the sample data. Most statistical computing packages produce an overall test of the model. This is a chi-square statistic (sometimes labelled the likelihood ratio chi-square) and tests whether the set of predictors is statistically significantly associated with time-to-event data. There are also measures of discrimination and calibration that are applicable to Cox proportional hazards regression models.[14-16]

In **EXAMPLE 3-12**, we use Cox proportional hazards regression to evaluate the association between electrical patterns in the heart and incident atrial fibrillation (AF) adjusted for age, sex, and other clinical risk factors.

EXAMPLE 3-12

Atrial fibrillation (AF) is a condition characterized by irregular heartbeat and was estimated to affect more than 33 million people worldwide in 2010.[17] There are several reported causes of AF, including damage to the heart's structure or electrical system, often due to other issues such as uncontrolled high blood pressure.[18] A sample of $n = 4{,}055$ participants, aged 45 years and older and free of AF, agree to participate in a 10-year prospective study to evaluate predictors of AF. At entry into the study, each participant undergoes an electrocardiogram (ECG) and electrophysiologists classify each participant's heart as having a normal, or an abnormal, electrical pattern.

Over the 10-year follow-up period, 17 of the 127 participants with an abnormal electrical pattern in their hearts develop AF compared to 215 of 3,928 participants with a normal electrical pattern in their hearts. We begin the analysis by first summarizing the cumulative incidence of AF over time between the groups in **FIGURE 3-4**.[f]

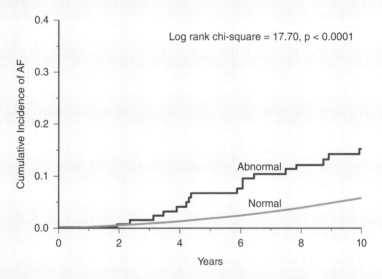

FIGURE 3-4 Cumulative incidence of atrial fibrillation over 10-year follow-up in patients with normal and abnormal electrical patterns in the heart.

---

f   Cumulative incidence is computed by subtracting the Kaplan-Meier estimates from 1, assuming no competing risks.

The cumulative incidence of AF is higher in participants with abnormal compared to normal electrical patterns in their hearts. The difference in cumulative incidence is statistically significant based on a logrank test as the logrank χ^2 statistic $= 17.70$, with $p < 0.0001$. We now explore whether there are other explanations for the observed difference in cumulative incidence using Cox proportional hazards regression analysis.

We use a statistical computing package to estimate three Cox proportional hazards regression models. The first is the crude, or unadjusted, model relating time-to-AF over 10 years to our indicator of abnormal electrical patterns in the heart. We then estimate a model that also includes age, in years, and an indicator of male sex to produce an age- and sex-adjusted estimate of the association between abnormal electrical activity in the heart and incident AF. The final model is a fully adjusted model that adjusts for age, sex, SBP, DBP, treatment for hypertension, and smoking status. The results are summarized in **TABLE 3-10**.

TABLE 3-10 Cox Proportional Hazards Regression Analysis Relating 10-Year Incident Atrial Fibrillation (AF) to an Indicator of Abnormal Electrical Activity in the Heart

Predictor	Regres- sion Coeffi- cient	Standard Error	χ^2	p	Hazard Ratio	95% Confi- dence Interval
Crude, Unadjusted Model						
Abnormal electrical activity	1.016	0.252	16.255	<0.0001	2.762	(1.685, 4.526)
Age- and Sex-Adjusted Model						
Age, years	0.083	0.006	167.069	<0.0001	1.087	(1.073, 1.100)
Male sex	0.647	0.133	23.536	<0.0001	1.910	(1.470, 2.480)
Abnormal electrical activity	0.367	0.256	2.056	0.1516	1.443	(0.874, 2.383)
Fully Adjusted Model						
Age, years	0.070	0.008	80.614	<0.0001	1.073	(1.056, 1.089)
Male sex	0.694	0.136	26.071	<0.0001	2.002	(1.534, 2.612)
Systolic blood pressure	0.010	0.004	7.438	0.0064	1.011	(1.003, 1.028)
Diastolic blood pressure	−0.010	0.008	1.792	0.1807	0.990	(0.975, 1.005)
Treatment for hypertension	0.452	0.141	10.275	0.0013	1.571	(1.192, 2.072)
Current smoker	0.088	0.152	0.339	0.5604	1.092	(0.811, 1.471)
Abnormal electrical activity	0.267	0.261	1.052	0.3050	1.306	(0.784, 2.177)

The statistical computing package produces an overall test for each model (a likelihood ratio test) and each model is statistically significant at $p < 0.0001$, which indicates that the respective sets of predictors are statistically significant.

In the crude, or unadjusted, model the estimated 10-year hazard or risk of AF is 2.762 times higher in participants with abnormal electrical patterns in their hearts, and this is statistically significant as the 95% CI does not include 1 (the null value for a hazard ratio) and $p < 0.0001$. The discrimination of the crude model is poor with a c statistic = 0.520. After adjustment for age and sex, the effect of an abnormal electrical pattern in the heart is no longer statistically significant as the hazard ratio is estimated at 1.443 with a 95% CI of (0.874, 2.383) and $p = 0.1516$. The discrimination of the model including age, sex, and the indicator of abnormal electrical pattern in the heart is much improved with a c statistic = 0.757, which is largely due to age and sex (and not to the indicator of abnormal electrical pattern in the heart). The magnitude of confounding due to age and sex is computed as $(2.762 − 1.443)/1.443 = 0.91$, which well exceeds the 10% rule. In fact, after adjustment for age and sex, the indicator of abnormal electrical pattern in the heart is no longer statistically significant. The fully adjusted model does not change the message, but after adjustment for age, sex, SBP, DBP, treatment for hypertension, and smoking status, the effect of abnormal electrical pattern in the heart on incident AF over 10 years is further diminished with an estimated hazard ratio of 1.306,which is not statistically significant as the 95% CI is (0.784, 2.177) and $p = 0.3050$. The other predictors in the fully adjusted model are interpreted as follows. Each additional year of age is associated with a 7.3% higher 10-year hazard or risk of AF (or 1.073 times the 10-year risk or hazard of AF), adjusting for sex, SBP, DBP, treatment for hypertension, smoking status, and abnormal electrical pattern in the heart. On average, males have twice the 10-year hazard or risk of AF compared to females, adjusting for age, SBP, DBP, treatment for hypertension, smoking status, and abnormal electrical pattern in the heart. Each additional 1 mmHg increase in SBP is associated with a 1.1% higher 10-year hazard or risk of AF (or 1.011 times) adjusting for age, sex, DBP, treatment for hypertension, smoking status, and abnormal electrical pattern in the heart. And patients on treatment for hypertension have 1.571 times the 10-year hazard or risk of AF compared to those not on treatment after adjusting for age, sex, SBP, DBP, smoking status, and abnormal electrical pattern in the heart.

▶ **3.7 Summary**

In many applications, we are interested in the associations between a primary risk factor or predictor and an outcome, but need to account for other risk factors that also play a role. In other applications, we aim to understand how, and to what extent, any of several risk factors or predictors are related to an outcome when considered simultaneously. Multivariable statistical analysis is a statistical approach to control for confounding (a distortion of the effect of a risk factor or predictor due to other variables) and to evaluate the relative importance of individual predictors on an outcome.

In this unit, we focused on the use of multivariable statistical analysis to control for confounding. It is important to recognize that there is always a potential for residual confounding, which refers to a distortion of the effect of a risk factor or predictor due to other variables that remains after statistical adjustment. Residual confounding might occur when there are other confounders that are not measured in the study, and thus, the investigator cannot control for them. As previously noted, it is always important to ask which other factors might explain any observed association, or lack thereof, when interpreting results, and residual confounding should be considered as a potential explanation.

Multivariable linear regression analysis is used to relate one or more risk factors or predictors to a single continuous outcome. Regression coefficients quantify the association between each risk factor or predictor and the outcome, after controlling for—or taking into account—the other predictors. Specifically, regression coefficients represent the expected change in the outcome relative to a 1-unit change in the predictor, adjusted for other variables considered. Risk factors or predictors can be tested for statistical significance based on t-tests and associated p-values, or by examining CI estimates for the regression coefficients.

Multivariable logistic regression analysis is used to relate one or more risk factors or predictors to a single dichotomous outcome. Regression coefficients quantify the association between each risk factor or predictor and the logit or log odds of the outcome, after controlling for—or taking into account—the other predictors. Specifically, regression coefficients represent the expected change in the log odds of the outcome relative to a 1-unit change in the predictor, adjusted for other variables considered. Usually, investigators exponentiate the regression coefficients to produce $\widehat{OR}$s which correspond to a 1-unit change in the predictor, adjusted for other variables considered. Risk factors or predictors can be tested for statistical significance based on χ^2 tests and associated p-values, by examining CI estimates for the regression coefficients, or by examining CI estimates for the ORs.

Time-to-event data are often summarized in survival curves. Differences in survival curves in two or more groups are tested using the logrank test and predictors of time-to-event outcomes are analyzed using Cox proportional hazards regression analysis. Regression coefficients quantify the association between each risk factor or predictor and the log hazard of the outcome, after controlling for—or taking into account—the other predictors. Specifically, regression coefficients represent the expected change in the log of the hazard at time, t, relative to a 1-unit change in the predictor, adjusted for other variables considered. Usually, investigators exponentiate the regression coefficients to produce estimates of hazard ratios which correspond to 1-unit changes in the predictor, adjusted for other variables considered. Risk factors or predictors can be tested for statistical significance based on χ^2 tests and associated p-values, by examining CI estimates for the regression coefficients or by examining CI estimates for the hazard ratios.

When evaluating multivariable models, it is important to assess model fit. A popular measure of model fit for multivariable linear regression is R^2, which quantifies the percentage of variation in the outcome explained by the predictors. Popular measures of model fit for multivariable logistic regression and Cox proportional hazards regression are measures of discrimination and calibration.

We now return to the population health issue we started this unit with—diabetes prevalence increasing worldwide—and apply some of the techniques we discussed to address each question.

- What are the modifiable risk factors for Type I and Type II diabetes?

  Predictors or risk factors can be classified as modifiable or nonmodifiable. Nonmodifiable risk factors include genetics, race/ethnicity, and biological sex assigned at birth. Modifiable risk factors are conditions or behaviors that can, in theory, be changed. We often focus on modifiable risk factors to design interventions to move individuals and communities toward healthier conditions or behaviors.

  Predictors of Type I diabetes include nonmodifiable risk factors such as genetics and race/ethnicity and also certain viral infections, environmental factors, and diet, all of which are potentially modifiable.[19] Predictors of Type II diabetes include genetics and race/ethnicity, which are nonmodifiable risk factors but also include obesity, high blood pressure, depression, low levels of high-density lipoprotein cholesterol, and high triglyceride levels.[20] These risk factors have been identified using multivariable logistic regression analysis relating prevalence or incidence of Type I diabetes or Type II diabetes, considered separately, to candidate risk factors. For example, a study by Yeung[21] reports that patients with enterovirus[g] had 9.8 times the odds of Type I diabetes with a 95% CI for the OR of (5.5, 17.4).[21] A study by Babu et al.[22] investigated the association between obesity and Type II diabetes in India, where the prevalence of obesity has been increasing steadily, and reported an OR of 1.14 with a 95% CI of (1.04, 1.24) (i.e., obese persons have 14% higher odds of Type II diabetes compared to nonobese persons).[22]

---

g Enteroviruses are viruses that cause infectious illness ranging from mild to more serious and are characterized by fever, respiratory issues, and other flu-like symptoms

■ Are there effective treatments for gestational diabetes to minimize the risks of adverse pregnancy outcomes?

Gestational diabetes affects approximately 25% of all pregnancies worldwide.[23] Women with gestational diabetes are at elevated risk for a number of adverse pregnancy outcomes including babies with very high birthweight (which can complicate delivery), premature birth, respiratory distress, and low blood sugar levels in the baby.[24] Controlling a pregnant mother's blood glucose level is important for the mother and the baby, and effective treatments include eating a healthy diet, engaging in physical activity, and, for some, insulin injections. However, there are some concerns about risks to the mother and the baby that might result from treatment, so there are not yet many large studies available to investigate the safety and efficacy of treatments for gestational diabetes. There was one randomized control trial of almost 1,000 pregnant mothers with mild gestational diabetes that found no statistical difference in risk of stillbirth, perinatal death, or neonatal complications for women receiving combination treatment of a dietary intervention, ongoing self-monitoring of blood glucose levels, and insulin therapy (if needed) compared to standard prenatal care.[25] More research is needed to fully evaluate whether different treatments can safely and effectively reduce risks of adverse pregnancy outcomes in women with gestational diabetes, and multivariable methods will play a key role in ensuring that investigations carefully control for confounding.

■ Are patients with Type II diabetes at higher risk of death due to cardiovascular disease?

Type II diabetes affects millions of people worldwide and its prevalence continues to rise. There have been many studies relating diabetes to incident cardiovascular disease and death due to cardiovascular disease, dating back decades. One of the largest, early reports was based on the Framingham Heart Study which reported a 2-fold higher risk of cardiovascular disease among men and women with diabetes compared to their nondiabetic counterparts, adjusted for age, systolic blood pressure, number of cigarettes smoked per day, and left ventricular hypertrophy using multivariable logistic regression analysis.[26] The Framingham Heart Study is one of the longest running, epidemiological studies of risk factors for cardiovascular disease.[27] Many risk factors or predictors for cardiovascular disease were first discovered in this landmark study. That said, the study participants were predominantly middle class and white, and thus results from the Framingham Heart Study do not necessarily generalize to other groups. Many other studies have since replicated many of the early findings in more diverse populations. For example, a recent Korean study reports a statistically significantly elevated risk of cardiovascular disease and cardiovascular disease death among participants with diabetes with hazard ratios of 1.87 [95% CI (1.38, 2.53)] and 4.10 [95% CI (2.20, 7.61)], respectively, after adjustment for age, sex, and cigarette smoking.[28]

The multivariable analysis procedures outlined in this unit are very flexible techniques that are widely used in practice to evaluate individual or joint effects of risk factors or predictors on outcomes. Here we focus on when to use which specific model and on the statistical interpretation of these models. Statistical significance is merely one way to interpret results, clinical or practical significance should always be evaluated. And last, there are many more details that need careful attention when constructing and interpreting multivariable models and interested readers should explore the references cited for more in-depth coverage of these technical and analytical details.

Key Points

- Randomization, restriction, and matching are popular methods to control for confounding in the design phase of a study but can be challenging to implement in certain scenarios.
- Multivariable statistical analysis is a popular statistical approach to control for confounding in the analysis phase of a study.
- Predictors in multivariable regression models can be continuous variables, dichotomous variables, or a set of dummy variables.
- Multivariable linear regression analysis is used to model a continuous outcome and the key measure of effect or association is the regression coefficient, which quantifies the expected change in the outcome relative to a 1-unit change in the predictor, adjusting for other variables in the model.
- Multivariable logistic regression analysis is used to model a dichotomous outcome and the key measure of effect or association is the odds ratio (OR), estimated by exponentiating the regression coefficient, which corresponds to a 1-unit change in the predictor, adjusting for other variables in the model.
- Cox proportional hazards regression analysis is used to model a time-to-event outcome and the key measure of effect or association is the hazard ratio, estimated by exponentiating the regression coefficient, which corresponds to a 1-unit change in the predictor, adjusting for other variables in the model.

References

1. American Diabetes Association. Complications. http://www.diabetes.org/living -with-diabetes/complications/. Accessed July 19, 2019.
2. March of Dimes. Gestational diabetes. https://www.marchofdimes.org/complications /gestational-diabetes.aspx. Accessed July 19, 2019.

3. Blotsky AL, Rahme E, Dahhou M, Nakhla M, Dasgupta K. Gestational diabetes associated with incident diabetes in childhood and youth: a retrospective cohort study. *CMAJ* 2019;191(15):E410-E417.

4. World Health Organization. Diabetes. https://www.who.int/news-room/fact-sheets /detail/diabetes. Accessed July 10, 2019.

5. Ascengrau A, Seage GR III. *Essentials of Epidemiology in Public Health.* 3rd ed. Burlington, MA: Jones & Bartlett Learning; 2014.

6. Armitage P, Berry G, Matthews JNS. *Statistical Methods in Medical Research.* 4th ed. Malden, MA: Blackwell Publishing Company; 2002.

7. Hosmer DW, Lemeshow S. *Applied Logistic Regression.* 2nd ed. New York, NY: John Wiley & Sons, Inc; 2000.

8. Steyerberg EW, Vickers AJ, Cook NR, et al. Assessing the performance of prediction models: a framework for some traditional and novel measures. *Epdemiology* 2010;21(1):128-138.

9. Kaplan EL, Meier P. Nonparametric estimation from incomplete observations. *J Am Stat Assoc.* 1958;53:457-481.

10. Rich JT, Neely JG, Paniello RC, Voelker CC, Nussenbaum B, Wang EW. A practical guide to understanding Kaplan-Meier curves. *Otolaryngol Head Neck Surg.* 2010;143(3):331-336.

11. Austin PC, Lee DS, Fine, JP. Introduction to the analysis of survival data in the presence of competing risks. *Circulation* 2016;133:601-609.

12. Bland JM, Altman, DG. The logrank test. *Br Med J.* 2004;328(7447):1073.

13. Corder GW, Foreman DI. *Nonparametric Statistics. A Step-by-Step Approach.* 2nd ed. Hoboken, NJ: John Wiley & Sons, Inc.; 2014.

14. D'Agostino RB, Nam B-H. *Handbook of Statistics. Evaluation of the Performance of Survival Models: Discrimination and Calibration.* Vol. 23. Amsterdam: Elsevier; 2003.

15. Pencina MJ, D'Agostino RB. Overall C as a measure of disrimination in survival analysis: model specific population value and confidence interval estimation. *Stat Med.* 2004;12:2109-2123.

16. Uno H, Cai T, Pencina MJ, D'Agostino RB, Wei LJ. On the C-statistics for evaluating overall adequacy of risk prediction procedures with censored survival data. *Stat Med.* 2011;30(10):1105-1117.

17. Chugh SS, Havmoeller R, Narayanan K, et al. Worldwide epidemiology of atrial fibrillation. A Global Burden of Disease 2010 Study. *Circulation* 2014;129(8):837-847.

18. American Heart Association. Why atrial fibrillation (AF or AFib) matters. https:// www.heart.org/en/health-topics/atrial-fibrillation/why-atrial-fibrillation-af-or -afib-matters. Accessed July 19, 2019.

19. Smith-Marsh DE. Type 1 diabetes risk factors. https://www.endocrineweb.com /conditions/type-1-diabetes/type-1-diabetes-risk-factors. Accessed July 23, 2019.

20. Risk factors for type 2 diabetes. https://www.niddk.nih.gov/health-information /diabetes/overview/risk-factors-type-2-diabetes. Accessed July 23, 2019.

21. Yeung W-CG, Rawlinson WD, Craig ME. Enterovirus infection and type I diabetes mellitus: systematic review and meta-analysis of observational molecular studies. *BMJ* 2011;342:d35.

22. Babu GR, Murthy GVS, Ana Y, et al. Association of obesity with hypertension and type 2 diabetes mellitus in India: a meta-analysis of observational studies. *World J Diabetes* 2018;9(1):40-52.

23. Immaneul J, Simmons D. Screening and treatment for early-onset gestational diabetes mellitus: a systematic review and meta-analysis. *Curr Diabetes Rep.* 2017;17(11):115.

24. Gestational diabetes. https://www.mayoclinic.org/diseases-conditions/gestational-diabetes/symptoms-causes/syc-20355339. Accessed July 24, 2019.

25. Landon MB, Spong CY, Thom E, et al. A multicenter, randomized trial of treatment for mild gestational diabetes. *NEJM* 2009;361:1339-1348.

26. Kannel WB, McGee DL. Diabetes and cardiovascular disease: the Framingham Study. *JAMA* 1979;241(19):2035-2038.

27. The Framingham Heart Study. https://www.framinghamheartstudy.org/fhs-about/history/epidemiological-background/. Accessed July 24, 2019.

28. Bae JC, Cho NH, Suh S, Kim JH, Jin S-M, Lee M-K. Cardiovascular disease incidence, mortality and case fatailty related to diabetes and metabolic syndrome: a community-based prospective study (Ansung-Ansan cohort 2001–2012). *J Diabetes* 2015;7:791-799.

Glossary

Adjusted Analysis an analysis that incorporates important variables that are associated with the outcome or are potential confounding variables (i.e., variables that are associated with the outcome and also with the risk factor of interest).

Alpha (α) the level of significance in a test of hypothesis. Defined as the probability of a Type I error, or the probability of rejecting the null hypothesis when, in fact, it is true.

Analysis of Variance (ANOVA) a popular procedure for testing the equality of k ($k > 2$) independent group means.

Bar Chart a graphical display for a categorical variable where the distinct response options are shown on the horizontal axis. Bars are centered over each response option with spaces in between adjacent responses, and the heights of the bars represent either the frequencies or relative frequencies of each response shown on the vertical axis.

Bayes' Rule A procedure for updating a prior probability by taking into account new information.

Beta (β) the probability of a Type II error in a test of hypothesis, or the probability of not rejecting the null hypothesis when, in fact, it is false.

Bias a systematic error in the design or analysis of a study that produces an estimate of association that exceeds or is smaller than the true association.

Binary Variable also called a dichotomous variable. A variable with exactly two possible responses (e.g., yes/no), usually coded 0 = no and 1 = yes, for analysis purposes.

Biostatistics the application of statistical principles to medical, public health, or biological problems.

Box Plot a graphical display of a continuous variable that shows the range of the data (minimum to maximum values), the first quartile, the median, and the third quartile.

Calibration a measure of how well predicted probabilities of outcome status, derived from a statistical model, agree with actual outcome status.

Categorical Variable also called a nominal variable. A variable with a fixed number of unordered response options.

Censoring this occurs in survival analysis when true survival time is not known because the study ends or because a participant drops out of the study before experiencing the outcome of interest. What is known is that the participant's survival time is greater than their last observed follow-up time, and these times are called censored times or censored data.

Central Limit Theorem a very important theorem in statistics that states that if simple random samples of size n are taken from a population with replacement, then for large samples (usually defined as samples with $n \geq 30$), the sampling distribution of the sample means is approximately normally distributed with a mean of μ and a standard deviation of $\sigma/\sqrt{n}$.

Chi-square Test of Independence a test of hypothesis used to assess whether there is a difference in the distribution of responses to a categorical or ordinal variable among independent comparison groups.

Chi-square (χ^2) Test Statistic a test statistic that follows the χ^2 distribution, which is positive and skewed.

Coefficient of Determination denoted R^2, this quantifies the percent of variability in the outcome that is explained by the predictor(s).

Column Percentage computed by dividing the cell frequency by the column total.

Competing Risk in survival or time-to-event analysis, another outcome that makes it impossible for the outcome of interest to occur.

Confidence Interval a range of plausible values for a population parameter with a level of confidence attached (e.g., 95% confidence).

Confidence Level the theoretical probability that a confidence interval will contain the true, unknown parameter. In practice, confidence levels of 90%, 95%, and 99% are used.

Confounding a distortion, exaggeration, or masking of an association between a risk factor and an outcome due to another variable.

Continuous Variable sometimes called a quantitative or measurement variable. Takes on an unlimited number of responses between a defined minimum and maximum value.

Correlation Coefficient also called the Pearson Product Moment correlation coefficient, denoted r. Ranges between -1 and $+1$ and quantifies the direction and strength of the linear association between two continuous variables.

Cox Proportional Hazards Regression a technique used in survival analysis to relate several risk factors or predictors, considered simultaneously, to time-to-event.

Cross-tabulation a table that summarizes frequencies according to two dichotomous, categorical, or ordinal variables.

Crude Analysis also called unadjusted analysis. Quantifies the association between a risk factor and an outcome without accounting for other variables.

Cumulative Incidence the ratio of the number of new cases of disease to the total number of participants who are at risk, or disease-free at the start of the study. In survival analysis, cumulative incidence is computed as 1 – the survival probability, and the latter can be computed using the Kaplan–Meier approach, assuming there are no competing risks.

Data–ink Ratio an assessment of the ink dedicated to data as opposed to other design features in a data visualization. The goal is to maximize the data–ink ratio.

Data Visualization a graphical display of data or statistical results.

Descriptive Statistics numerical or graphical summaries of data collected in a sample.

Diagnostic Test a test used to establish presence or absence of disease.

Dichotomous Variable also called a binary variable. A variable with exactly two possible responses (e.g., yes/no), usually coded 0 = no and 1 = yes, for analysis purposes.

Discrimination a measure of how well a statistical model distinguishes between participants with versus those without the outcome of interest.

Distribution the collection of responses to a particular measurement in a sample or group.

Dummy Variables a set of binary or dichotomous variables used to differentiate among more than two levels of a categorical or grouping variable.

Effect Modification a situation that occurs when there is a different relationship between the exposure or risk factor and the outcome, depending on the level of another characteristic or variable. Also called statistical interaction.

Equivalence Test a test of hypothesis where the goal is to demonstrate equality of outcomes, as opposed to a difference in outcomes, requiring a different statistical approach to hypothesis testing than that discussed in this book.

Estimation the process of determining plausible values for a population parameter based on a random sample.

Expected Frequency the number of participants that would be expected in a group under certain assumptions. Expected frequencies are part of χ^2 tests, where they are computed based on the assumption that the null hypothesis is true.

False Negative Fraction the probability that a diseased participant tests negative.

False Positive Fraction the probability that a disease-free participant tests positive.

First Quartile denoted Q_1, the value in the dataset that separates the bottom 25% of the values from the rest. Equivalent to the 25th percentile.

Frequency the number of participants in a particular group.

Generalizability in biostatistics, results observed in a sample are said to be generalizable to the population as long as the sample is truly representative of that population.

Hazard the instantaneous risk of experiencing an event.

Hazard Ratio the ratio of hazards in two independent comparison groups, interpreted as the risk of event in the group in the numerator as compared to the risk of event in the group in the denominator at a specific time.

Histogram a graphical display for an ordinal variable where the distinct response options are shown on the horizontal axis. Bars are centered over each response option, and the heights of the bars represent either the frequencies or relative frequencies of each response shown on the vertical axis. There is no gap between response options on the horizontal axis, suggestive of the underlying continuum in ordered response options.

Historical Control a group of participants analyzed or treated in the past that are used as a comparator in a current study.

Hypothesis Testing a process whereby a specific statement or research hypothesis is generated about a population parameter and compared to a null hypothesis that reflects the no difference or no association scenario and sample statistics are evaluated to assess the likelihood that the research hypothesis is true.

Incidence also called incidence proportion or risk. Computed as the ratio of the number of new cases (e.g., of disease) over a period of time to the total number at risk.

Incidence Rate Computed as the ratio of the number of new cases of disease to the total follow-up time (i.e., the sum of all disease-free person-time).

Information Bias systematic errors in the measurement of risk factors or outcomes or in the classification of participants in terms of risk factors or outcomes.

Interaction also called statistical interaction or effect modification. Occurs when there is a different relationship between the exposure or risk factor and the outcome depending on the level of another characteristic or variable.

Intercept also called the y-intercept. In linear regression analysis, the value of the outcome variable (y) when the predictor variable(s) are all equal to zero.

Interquartile Range a measure of variability, computed as the difference between the first and third quartiles, interquartile range = $Q_3 - Q_1$.

Kaplan–Meier Method also called the product-limit approach. A popular approach for summarizing the survival experiences of participants over a pre-defined follow-up period until the time they experience the event or outcome of interest or until the end of the study, whichever comes first. In this approach, the survival probability is re-estimated each time an event occurs (as opposed to using equally spaced intervals, as in the life table or actuarial approach).

Kruskal–Wallis Test a nonparametric test used to compare medians among k independent comparison groups ($k > 2$), sometimes described as an ANOVA with the data replaced by ranks.

Level of Significance denoted α. Used in a test of hypothesis and defined as the probability of a Type I error, or the probability of rejecting the null hypothesis when, in fact, it is true.

Life Table also called an actuarial table and used to summarize the survival experiences of participants over a predefined follow-up period until the time they experience the event or outcome of interest, or until the end of the study, whichever comes first.

Linear Regression Analysis a technique used to estimate the association between one or more predictors or risk factors and a single continuous outcome.

Logistic Regression Analysis a technique used to estimate the association between one or more predictors or risk factors and a single dichotomous outcome.

Logit also called the log odds. Used in logistic regression analysis and computed by taking the log of the odds of experiencing the outcome of interest.

Logrank Test a nonparametric test to test the null hypothesis of no difference in survival among two or more independent groups.

Mann–Whitney U Test sometimes called the Mann–Whitney–Wilcoxon test or the Wilcoxon Rank Sum test. A popular nonparametric test used to compare outcomes between two independent groups to determine whether two samples are likely to derive from the same population, or that the two populations have the same shape.

Margin of Error the product of the value that reflects the desired confidence level (e.g., $z = 1.96$ for 95% confidence) and the standard error of the point estimate.

Matched Samples also called paired samples, where observations within each matched pair are related. This relationship must be reconciled in statistical analysis.

Matching a process used to control confounding by organizing comparison groups on the basis of similar characteristics.

Maximum the largest value.

Maximum Likelihood Estimation an optimization technique used to estimate regression coefficients in logistic regression models.

McNemar's Test a test for the equality of proportions when the samples are matched or paired, sometimes called McNemar's test for dependent proportions.

Mean a measure of central tendency or location, computed as the ratio of the sum of a set of values to its size (e.g., the population mean is $\mu = \dfrac{\sum x}{N}$ and the sample mean is $\bar{X} = \dfrac{\sum x}{n}$).

Measurement Variable sometimes called a continuous or quantitative variable. Takes on an unlimited number of responses between a defined minimum and maximum value.

Median a measure of central tendency or location, defined as the value that separates the top 50% of values from the bottom 50%, sometimes called the middle value or the 50th percentile.

Minimum the smallest value.

Mode the most frequently occurring value in a set of observations.

Multiple Comparison Procedures sometimes called post-hoc tests, statistical procedures that control the overall Type I error over a series of hypothesis tests (e.g., many pairwise tests) following a statistically significant overall test.

Multivariable Analysis analyses designed to assess relationships among several risk factors or exposures and a single outcome.

Multivariable Linear Regression Analysis a technique used to estimate associations between two or more predictors or risk factors and a single continuous outcome.

Multivariable Logistic Regression Analysis a technique used to estimate associations between two or more predictors or risk factors and a single dichotomous outcome.

Negative Confounding when the observed, crude association is masked or biased toward the null hypothesis.

Negative Predictive Value the probability that a participant who tests negative is disease-free.

Nominal Variable also called a categorical variable. A variable with a fixed number of unordered response options.

Non-inferiority Test a test of hypothesis where the goal is to demonstrate that outcomes in one treatment or group are not much different from those in another, requiring a different statistical approach to hypothesis testing than that discussed in this book.

Nonparametric Tests sometimes called distribution-free tests as they are based on few assumptions (e.g., they do not assume that the outcome follows a particular distribution).

Normal Probability Distribution a probability distribution for a continuous outcome that is well described by a bell-shaped curve. In a normal distribution, mean = median = mode, and the distribution is symmetric about the mean. In addition, approximately 68% of the values fall between the mean and one standard deviation in either direction (i.e., $\mu - \sigma < X < \mu + \sigma$, where μ is the population mean and σ is the population standard deviation); approximately 95% of the values fall between the mean and two standard deviations in either direction (i.e., $\mu - 2\sigma < X < \mu + 2\sigma$); and approximately 99.9% of the values fall between the mean and three standard deviations in either direction (i.e., $\mu - 3\sigma < X < \mu + 3\sigma$).

Null Hypothesis statement of no difference, no association, or no effect. Tested against a research hypothesis, which postulates a difference, an association, or an effect.

Odds the ratio of the number of participants with the outcome to those without.

Odds Ratio the ratio of two odds. For example, the ratio of the odds of disease in an exposed group compared to the odds of disease in an unexposed group.

One-sided Test a test of hypothesis in which investigators of the research hypothesis reject the null hypothesis if the test statistic is extreme in a particular direction.

Ordinal Variable a variable with a fixed number of ordered response options.

Ordinary Least Squares an estimation technique that minimizes the sum of squared residuals to produce estimates of linear regression coefficients.

Outcome the primary response variable in an analysis.

Outliers values outside the range of those generally expected. A popular guideline to detect outliers is as follows: Outliers are values below $Q_1 - 1.5$ IQR or above $Q_3 + 1.5$ IQR, where IQR is the interquartile range.

p-**value** the exact statistical significance of the data, or the incompatibility of the sample data with the assumed statistical model.

Paired Samples also called matched samples, where observations within each matched pair are related. This relationship must be reconciled in statistical analysis.

Paired *t* Test a test for the equality of means in two matched or paired samples, based on an analysis of difference scores.

Parameter any summary measure computed on a population.

Parametric Test a test of hypothesis where the outcome is assumed to follow a particular probability distribution (e.g., the normal distribution) and where the test involves estimation of the key parameters of that distribution (e.g., the mean or difference in means) based on sample data.

Pearson Product Moment Correlation Coefficient denoted *r*, ranges between −1 and +1 and quantifies the direction and strength of the linear association between two continuous variables.

Person-time a measure of time (e.g., in days or years) that participants in a study are at risk for a particular event or outcome of interest.

Point Estimate a single-valued estimate of a population parameter derived from a sample.

Population the collection of all individuals about whom investigators are interested in making generalizations or inferences.

Population Size the number of participants in the population, denoted *N*.

Positive Confounding when the observed, crude association is exaggerated or biased away from the null hypothesis.

Positive Predictive Value the probability that a participant who tests positive has the disease.

Posttest Probability also called a posterior probability, an estimate of the probability that a participant has an outcome of interest computed after taking additional evidence (e.g., the result of a screening test) into account.

Predictor a variable hypothesized to be associated with the outcome of interest.

Pretest Probability also called a prior probability, an estimate of the probability that a participant has an outcome of interest without taking additional evidence (e.g., the result of a screening test) into account.

Prevalence the proportion of individuals with a particular condition (e.g., disease) at a point in time.

Probability a number that reflects the likelihood that a particular outcome occurs.

Probability Model a mathematical equation or formula used to generate probabilities based on certain assumptions about a process.

Quantitative Variable sometimes called a continuous or measurement variable. Takes on an unlimited number of responses between a defined minimum and maximum value.

Quartiles values that divide a distribution into four equally sized groups.

R^2 also called the coefficient of determination. Quantifies the percent of variability in the outcome that is explained by the predictor(s).

Randomization a process by which participants are assigned by chance to different treatment or exposure groups.

Range a measure of variability, computed by taking the difference between the minimum and maximum values, range = maximum – minimum.

Rate the likelihood that an individual changes status (e.g., develops disease) in a specified unit of time.

Ratio computed by dividing one quantity by another, and the numerator and denominator need not be related.

Receiver Operating Characteristic (ROC) Curve a visual display of sensitivity on the y-axis and the false positive fraction on the x-axis.

Regression Analysis a technique used to relate one or more risk factors or predictors to an outcome.

Regression Coefficient an estimate of a parameter that quantifies the association between a risk factor or predictor and an outcome in a regression setting; it is interpreted as the change in the outcome relative to a one-unit change in the predictor or risk factor, adjusting for, or holding all other predictors, constant.

Relative Frequency the ratio of the number of participants in a particular group (frequency) to the total sample size.

Relative Risk also called risk ratio. The ratio of prevalence or incidence in the exposed group to the prevalence or incidence in the unexposed group.

Repeated Measures ANOVA a procedure for testing the equality of k ($k > 2$) related group means (e.g., multiple assessments measured serially in time on each participant).

Research Hypothesis represents the investigator's belief in terms of the association or effect under study.

Residual Confounding a distortion, exaggeration, or masking of an association between a risk factor and an outcome that remains after controlling for other variables either in the design or analysis of a study (e.g., because additional confounding variables were not measured or reconciled).

Restriction used to control confounding by narrowly defining sample eligibility criteria.

Right Censoring occurs when a participant does not have the event of interest during the study, and thus his or her last observed follow-up time is less than the time-to-event. This can occur when a participant drops out before a study ends or when a participant is event-free at the end of the observation period.

Risk Factor a characteristic or behavior that is hypothesized to be associated with the outcome of interest.

Risk Ratio also called relative risk. The ratio of prevalence or incidence in the exposed group to the prevalence or incidence in the unexposed group.

Row Percentage computed by dividing the cell frequency by the row total.

Sample a subset of the population, with participants in the sample ideally selected at random and representative of the population.

Sample Correlation Coefficient also called the Pearson Product Moment correlation coefficient. Denoted r, ranges between -1 and $+1$, and quantifies the direction and strength of the linear association between two continuous variables.

Sample Mean a measure of central tendency in a sample, computed as the ratio of the sum of the values in the sample to the sample size, $\bar{X} = \dfrac{\sum x}{n}$.

Sample Median a measure of central tendency in a sample reflecting the value that separates the top 50% of the values from the bottom 50%. Sometimes called the middle value or the 50th percentile of the sample.

Sample Mode the most frequently occurring value in a sample. A sample can have no mode (when each value occurs as many times as every other) or more than one mode (e.g., a bimodal sample).

Sample Proportion denoted $\hat{p}$ and computed by taking the ratio of the number of participants with a particular outcome or attribute in the sample to the sample size, $\left(\hat{p} = \dfrac{\text{number with outcome}}{n}\right)$.

Sample Range a measure of variability in a sample, computed by taking the difference between the minimum and maximum values in the sample.

Sample Size the number of participants in the study sample, denoted n.

Sample Standard Deviation a measure of variability in a sample, denoted s and computed as follows:

$$s = \sqrt{\frac{\sum (\bar{x} - x)^2}{(n-1)}}$$

Sampling a procedure by which individuals are selected from a population into a sample.

Sampling Distribution the probability distribution of a statistic produced by repeatedly selecting samples of the same size and computing the desired statistic (e.g., sampling distribution of the sample mean).

Scatter Plot a graphical display in which each point represents an (x, y) pair measured in a different participant. Scatter diagrams often display the predictor or risk factor (x) on the horizontal axis and the outcome (y) on the vertical axis.

Screening Test a test used for early detection of disease in people yet to experience signs or symptoms of disease.

Selection Bias a distortion of the association (an over- or underestimation of the true association) between an exposure or risk factor and an outcome resulting from the process of selecting or retaining participants in the study.

Semiparametric Method an approach with parametric and nonparametric components.

Sensitivity the true positive fraction or the probability that a diseased participant tests positive.

Sign Test a nonparametric test to compare outcomes between two matched or paired groups based on the numbers of positive and negative difference scores computed in matched pairs.

Simple Linear Regression Analysis a technique used to estimate the equation of the line that best describes the association between one predictor or risk factor and a single continuous outcome.

Simple Logistic Regression Analysis a technique used to estimate the equation that best describes the association between one predictor or risk factor and a single dichotomous outcome.

Simple Random Sample a procedure whereby individuals are selected at random from a population into a sample.

Slope in linear regression analysis, the change in the outcome (y) associated with a one-unit change in the predictor or independent variable (x).

Specificity the true negative fraction or the probability that a disease-free participant tests negative.

Standard Error the standard deviation of a summary measure or statistic.

Standard Normal Distribution a normal distribution with a mean of zero $(\mu = 0)$ and a standard deviation of one $(\sigma = 1)$.

Statistic any summary number computed in a sample.

Statistical Adjustment a procedure used to control confounding using multivariable analysis to account for relationships among variables.

Statistical Inference procedures used to make generalizations about a population based on analysis of sample data.

Statistical Interaction also called interaction or effect modification. Occurs when there is a different relationship between the exposure or risk factor and the outcome depending on the level of another characteristic or variable.

Statistical Significance the incompatibility of the sample data with the assumed statistical model.

Stratification a process whereby participants are partitioned or separated into mutually exclusive or non-overlapping groups.

Study Design the methodology used to collect the information from the study sample to address the research question.

Survival Analysis statistical methods and procedures for time-to-event outcomes.

Survival Function probabilities that participants survive past certain time points as a function of time.

t Distribution a probability model for a continuous variable that is similar to the standard normal distribution z, but takes a slightly different shape depending on the exact sample size. When the sample size is large, the t distribution is very similar to the standard normal distribution z.

t Statistic a test statistic that follows the t distribution.

Test Statistic a single number that summarizes important information in the sample and is used for hypothesis testing.

Third Quartile denoted Q_3. The value in a dataset that separates the top 25% of the values from the rest, equivalent to the 75th percentile.

Threshold of positivity a value such that if a participant measures at or above (or at or below, depending on the measure) that value, they are considered test positive.

Time-to-Event Variable a variable that reflects the time until a participant experiences an event of interest; used in survival analysis.

Time-varying predictor a predictor that changes over time during a study.

Two-factor ANOVA a procedure used to test for differences in means due to one or both grouping variables (called factors) or the statistical interaction between the two.

Two-sided Test a test in which the investigators of the research hypothesis reject the null hypothesis if the test statistic is extreme in either direction.

Type I Error the situation in which the null hypothesis is rejected when, in fact, it is true.

Type II Error the situation in which the null hypothesis is not rejected when, in fact, it is false.

Unadjusted Analysis also called crude analysis. Quantifies the association between a risk factor and an outcome without accounting for other variables.

Unbiased estimate an estimate that is expected, on average, to equal the true parameter.

Uniform Probability Distribution a distribution where each value of a continuous measure is equally likely to occur.

Variable any phenomenon, characteristic, or attribute that varies among individuals.

Wilcoxon Signed Rank Test a nonparametric test used to compare outcomes between two matched or paired groups that is based on the ranks of the absolute values of the difference scores computed in the matched pairs.

y-intercept in linear regression analysis, the value of the outcome variable (y) when the predictor or risk factor (x) is equal to zero.

z Score also called a z-value, computed as $z = \dfrac{X - \mu}{\sigma}$, where μ is the mean and σ is the standard deviation of a variable X that is assumed to follow a normal distribution.

Index

Note: "Page numbers followed by *f*, or *t* indicate material in boxes, figures, or tables, respectively."